The Ultimate Book of
Historic Barns

The Ultimate Book of
Historic Barns

HISTORY • LEGEND • LORE • FORM • FUNCTION • SYMBOLISM • ROMANCE

Robin Langley Sommer

THUNDER BAY
P·R·E·S·S

Published in the United States by
Thunder Bay Press
An imprint of the Advantage Publishers Group
5880 Oberlin Drive, Suite 400
San Diego, CA 92121-4794
http://www.advantagebooksonline.com

Produced by Saraband Inc., PO Box 0032, Rowayton,
CT 06853-0032

Copyright © 2000 Saraband Inc.

Design © Ziga Design

ISBN 1-57145-223-0

Library of Congress Cataloging-in-Publication Data

Sommer, Robin Langley.
    The ultimate book of historic barns / Robin Langley Sommer.
        p. cm.
    Includes bibliographical references and index.
    ISBN 1-57145-223-0
    1. Barns--United States--History. 2. Barns--United States--
History--Pictorial works. I.
Title

NA8230 .S63 2000
728'.922'0973--dc21   99-059348

10 9 8 7 6 5 4 3 2 1

These pages: Lancaster County, Pennsylvania
Page 1 photograph: North Carleton, Prince Edward Island
Page 2 photograph: Fremont County, Idaho
Page 3 photograph: Cupola detail, Mountain Valley Farm, Waitsfield, Vermont

For Eric David Sommer, Number One Son

The publisher would like to thank the following individuals for their assistance in the preparation of this book: Nicola J. Gillies and Sara Hunt, edi-
tors; Charles J. Ziga, art director; Wendy Ciaccia Eurell and Nikki L.
Fesak, graphic designers; Lisa Langone Desautels, indexer; and the pho-
tographers listed below. Grateful acknowledgement is also made to Marty
and Lone Azola, for permission to feature their home; Marilyn Holnsteiner,
for supplementary research; the numerous farmers and barn owners who
gave their permission and assistance in the photography of their property
(permission was sought and obtained wherever possible); and the travel
and tourism agencies and historical societies across the continent.
    Particular thanks are due to the photographers and agencies listed
below for permission to reproduce the photographs on the following
pages: © **Larry Angier:** 2, 181, 182–83, 184, 188–89, 190, 192, 193, 205,
206, 207, 214, 215, 217, 224b; © **Tony Arruza:** 88, 94, 96–97, 104–105;
© **2000 Kindra Clineff:** 23, 24, 26, 27, 32, 34, 36–37, 42–43, 82–83b, 220;
© **Robert Drapala:** 48, 56, 57, 244; © **Carolyn Fox:** 185, 186b, 191, 196,
204, 238; © **Anne Gummerson:** 100–101; © **Rudi Holnsteiner:** 79, 81,
147, 150–51, 152–53 (both), 157, 158–59t, 160–61 (both), 165, 168–69
(both), 170–71t, 172–73, 176, 178, 186–87t, 198–99 (both), 208–209,
226, 227, 230, 236–237tr; © **Karlene Kingery:** 140, 144, 162, 171b;
© **Balthazar Korab:** 62–63t, 64–65, 142, 163, 223, 232, 239; © **Rod
Patterson:** 30–31, 74–75, 77, 92–93, 95, 194–95 (both); © **Chuck Place:**
40, 174–75, 197; © **John Sylvester:** 1, 6–7, 106–107, 108, 112, 113, 116,
117, 118–19, 120, 121, 122–23, 124, 125, 126–27, 128, 129, 130, 131,
132–33, 134–35, 136–37, 202–203, 224–25t, 231, 235, 240–41, 242, 243,
252, 253; © **2000 Charles J. Ziga:** 3, 4–5, 14–15, 17, 19, 20, 25, 28–29
(both), 33, 35, 38–39 (both), 41, 44–45, 46, 47, 50, 51, 52, 53, 54–55, 58,
59, 60–61, 62b, 66, 67, 68–69, 70–71 (both), 72–73, 78, 83t, 84–85 (both),
86–87, 89, 90, 91, 98–99 (both), 102–103 (both), 138–39, 146, 149, 156,
159b, 164, 166–67 (both), 200, 201, 212, 213, 216, 218, 219, 221, 222,
228–29, 233, 234, 236 tl and bl, 245, 246 (both), 247, 248, 249 (both), 250
(both), 251; **Canadian Tourism Commission Photo:** 110 (Alan
Carruthers), 115t (John Devisser), 154 (Yves Beaulieu); **Lehigh County
Historical Society:** 8; **Library of Congress, Prints and Photographs
Division:** 10; **Montana Historical Society, Helena:** 210; **Ontario
Ministry of Economic Development, Trade & Tourism:** 114 (both), 115;
Provincial Archives of New Brunswick: 9; **Saskatchewan Archives
Board:** 13; **Saskatchewan Government Heritage Branch:** 148 (Frank
Korvemaker), 155 (Wayne Zelmer); **Sharlot Hall Museum
Library/Archives, Prescott, Arizona:** 11, 211; **South Dakota State
Historical Society—State Archives:** 12.

Contents

Introduction: Origins

Previous pages: Evening milking, Prince Edward Island; *Above:* Amish farm, Pennsylvania

As the English cultural historian John Ruskin wrote in *The Seven Lamps of Architecture* (1849), "Speaking of old buildings, they are not just ours. They partly belong to those who built them and partly to all generations of mankind who are to follow us." Today, this observation speaks to us of a unique achievement in vernacular architecture: the historic barns of North America, many of which have already vanished from our landscapes. Fortunately, others have been maintained or restored, some as working buildings and many as pioneer and agricultural museums, where city children who have never seen a barn may be captivated by a fragrant haymow, or a flock of sheep, as were many of us who grew up when summer's highlight was a long visit to an aunt and uncle's farm, or to an old family homestead where an abandoned quarry had filled up with spring water to make the perfect swimming hole.

The word barn comes from the Anglo-Saxon *berern*, derived from the words *bere* (barley) and *aern* (close place). The following pages provide an overview of the several European prototypes of American and Canadian barns and the adaptations made to them in the New World over a period of some four centuries. The principal innovation was to combine the European granary, used to store cereal crops, and the livestock shelter, formerly a separate facility, into a single building. During the early 1600s, English colonists in coastal Massachusetts and Virginia stored their first meager harvests in covered holes and sheltered their animals in makeshift lean-tos of pole frameworks covered with hay. However, the severity of the New England winters as compared to those of the Mother Country dictated sturdier shelters, sometimes connected to the dwelling and other outbuildings. Similarly, the original French settlers of Quebec found that stone barns were unsuitable in New France, as inside winter temperatures often dropped far below zero. Their timber-framed and log barns were modeled on the connected architecture of the Norman and Breton farmstead. Later, British Canadians in the future Maritime Provinces would build saltbox barns like those of their New England neighbors, with the principal roof slope facing north, "into the weather," and banked with hay or other insulating materials.

From the time when oxen were first used for plowing, the width of their stalls became a factor in the dimensions of European barns. The width of an ox stall was usually 4 feet; a "short yoke" of two oxen required 8 feet of stall room, and a "long yoke" of four animals, 16 feet. Thus the length of the barn tended to be some multiple of 8 or 16 feet, and 16 feet became the normal length for boards when they were hewn or sawed by hand. Barns were readily enlarged by building a lean-to along one or both sides and another across the rear. The barn then resembles the basilica plan of a medieval church, with nave, side aisles, choir (used as a dwelling space in the central European and Dutch house/barn) and clerestory—bands of windows set high in the walls. Such barns had also been used by medieval landholders and churchmen to store their tithes of grain.

The structural parts used in timber-frame construction until the mid-nineteenth century varied little from their medieval antecedents. They included a sill at the base of the building, often laid on a fieldstone foundation, and a series of "verticals," or upright posts formed from squared-off tree trunks. The space between each pair of posts was called a bay, and the traditional New England-style barn usually consisted of three bays used for various functions. The central portion was the threshing floor, when grain was still threshed by hand with flails, and the other two might house livestock and implements, respectively. A cattle barn would have an overhead hay bay, from which hay was dropped for the animals' feed. A shed added on later would constitute an extended bay.

The timber over the main entrance was called a door header, and cross-sections called bents were put together on the ground and raised in sections between the sill and the top timbers, or plates. Braces of diagonal timbers strengthened the structure and helped to support the roof. Joinery consisted of mortises cut into the sill to receive the tenons shaped at the bottom of

Saltbox barn-raising, New Brunswick

each corner post. These joints were secured by wooden pegs or "trunnels" (tree nails) that were fashioned by hand. Then the preconstructed bents were raised by long poles, sometimes with the help of ropes, and locked into place by the upper plates.

Rafters for the roof were joined at the peak by the same mortise-and-tenon method; later, the rafters were fastened to a ridgepole that ran the length of the roof. This became possible from about 1790, when crude machine-made iron nails became more widely available. Spaces between the frames were filled with woven sticks and mud (wattle-and-daub), often plastered over, or, among some ethnic groups, with rough stone or brickwork. For additional weatherproofing, the entire barn was often sheathed with horizontal or vertical boards. This applied to the log buildings introduced by French, German and Swedish settlers as well. Over time, many were clad in boards, or even plastered over.

Although many people associate the log cabin and barn with the original frontier, the continent's first buildings, from Upper Canada to the English colony of Virginia, were of timber-frame construction covered with hand-made clapboards and roofed with thatch. Eventually, thatching would be replaced by slabs of bark, boards, handmade shingles and, occasionally, sod, as seen in the oldest barns of Scandinavian origin.

The Benjamin Abbot farm complex in Andover, Massachusetts (founded in 1685), expanded through the years, like so many farmsteads, from its early form in 1685. The original house was connected to the barn by a woodshed. Random-width horizontal siding and clapboard covered both barn and shed. Eventually, the barn acquired a large hayloft and a workshop, and the ground floor housed wagons and livestock.

Shortly after the Revolutionary War, the Worcester, Massachusetts, *Gazette* featured the following

Above: Apple picking, West Virginia; *Opposite:* Hand-feeding orphaned piglets

advertisement: "A Farm of about 220 acres of land, lying in Sunderland, in the County of Hampshire, about one mile and an half from Connecticut River. The price will be low, and the conditions of payment made easy. For particulars, inquire of SAMUEL WARD, of Lancaster, in the county of Worcester." The prosperous state of New England farming was also reflected in an 1801 advertisement published in Chesterfield, New Hampshire: "FOR SALE…About 12 acres of good land, with a convenient Dwelling-house and a large Store adjoining a Barn, Shed, &c., all new and well finished. EBEN STEARNS."

Well before this time, Dutch settlers had built their distinctive gable-entry barns in the beautiful Hudson River Valley and in nearby New Jersey, Long Island and Staten Island. Their outbuildings included tall, narrow smokehouses of brick and stone in which the lower level contained the slow-burning fire used to smoke the meat hung and stored in the upper levels. Some of these smokehouses, like those found on the Pennsylvania German farms, were built of squared logs with diagonal braces. In Colonial Williamsburg, Virginia, an eighteenth-century English smokehouse from the Archibald Blair property has been restored from brick foundation to shingled roof. Smokehouses, outdoor ovens and the blacksmith's quarters were always located away from other farm buildings because of the danger of fire.

In the Delaware Valley and the Southeast, Swedish and German settlers built log "cribs" that were chinked with mud or moss as their first barns and dwellings. The Appalachian crib barn consisted of an open driveway between two—sometimes four—of these enclosures spanned by a common roof. A loft for hay storage might be added to form an upper level. One crib was generally used to shelter cows, the other, horses or mules. This region is also studded with many historic gristmills that used water power from streams and rivers to grind grain. The weathered Glade Creek Gristmill in West Virginia's Babcock State Park adds to the rustic beauty of this scenic state.

Inevitably, the first North American frontier expanded west across the Appalachians, pioneered by woodsmen like Daniel Boone, who discovered the Cumberland Gap

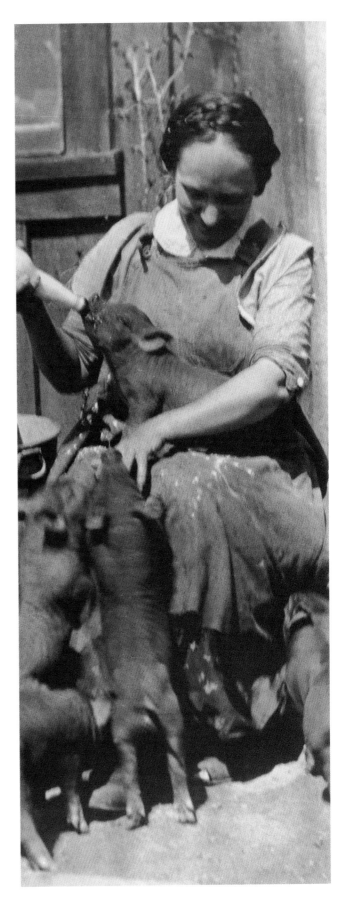

that opened the Midwest and the Great Plains to American settlement. As he told his biographer, "It was on the first of May, in the year 1769, that I resigned my domestic happiness for a time, and left my family and peaceable habitation on the Yadkin River, in North Carolina, to wander through the wilderness of America, in quest of the country of Kentucke." Fifty years later, Jedediah Smith and his fellow trappers found the South Pass that provided a route for wagons across the Rocky Mountains through to Oregon and California. The great westward migration of the nineteenth century was underway across the continent, and new types of shelter were devised to meet changing climates, crops and available materials. In Pennsylvania's Susquehanna Valley, the little hamlet of Mechanicsburg became a stopping place for travelers to repair their wagons, and second-generation German settlers took their ample banked barns into Ohio, Indiana and Iowa. In fact, the Pennsylvania-style barn would be imitated from western Ontario to Washington State.

Early settlers of the present-day Cornbelt included members of the Amana Society, who arrived in Iowa in 1855. The group had originated in Germany during the Protestant Reformation, as one of the many Pietist sects, then called the Community of True Inspiration. The name "Amana" is a variant of Abana, the Biblical name of Syria's River Barada. Centered in Hesse, Germany, the "Inspirationists," like the several

Mennonite sects, had suffered periodic persecution, and in 1843, Christian Metz led the first group to the United States, where they founded an agricultural commune near Buffalo, New York. Thence they moved to the western frontier—some 25,000 acres of rich Iowa farmland, where they established the first of their seven villages, or "colonies," named Old Amana. They built German-style barns to house their crops and livestock and also engaged in woolen manufacture and compounded herbal remedies, like the Shaker communities. The Amanites were widely respected, and their picturesque mill stream and well-kept villages attracted many visitors. Amana was considered one of the most successful of the many religious communes of its day. Not until the early twentieth century was it reorganized along the lines of a profit-sharing cooperative that became best known as a manufacturer of appliances.

The Great Plains of both the United States and Canada attracted pioneers from many European countries, who used the newly developed steel plow to cut through the tough prairie grass and built dugout homes and barns roofed with sod. Soon, grain fields of vast extent began to cover the windswept grasslands, which were dotted with prairie barns and windmills. Western farmers and ranchers experimented with round and octagonal barns like those pioneered by the Shakers in New England after the Civil War. Their famous round barn at Hancock, Massachusetts, was unique

Above: Early threshing machine, South Dakota; *Opposite:* Timber for a Saskatchewan prairie barn

for its size and the ideal arrangement of its various functions, but few farmers had the dedicated labor force and the necessary financial resources to duplicate this landmark dairy barn of stone. Most adaptations were of wood framing, and multisided rather than circular. As Byron D. Halsted pointed out in his 1881 book *Barn Plans and Outbuildings*, "There is no economy in building a strictly round barn, as curved walls, sills, cornice and roofing are very expensive and offset the trifling gain in floor space."

During the Cattle Kingdom days of the late 1800s, ranchers experimented with new breeds of beef cattle. Historically, the nation's major cattle breeds have derived from British and Indian stock, including the English Devon, Hereford, Red Poll, Shorthorn and Sussex; the Scottish Aberdeen Angus, Highland and Galloway; and the hump-backed Zebu cattle from India, which Americans collectively call "Brahman." Zebus have also been crossed with other stock because of their high resistance to hot weather and sparse grazing land. The new fixed breeds resulting from these crosses include the Santa Gertrudis, Beefmaster and Brangus. As Halsted pointed out in *Barn Plans and Outbuildings*, "Farmers in the newer portions of the West do not have stables for their cattle or snug sheds for their sheep....Sheds of poles with roofs of straw are extensively used and with profit. They furnish at the

same time shelter from storms and feed for the protected animals. New hay is packed on after each storm."

Western Canada, settled late by a diverse population from Europe, the United States and Eastern Canada, may serve as a microcosm of barn types and functions developed in the New World. As Bob Hainstock notes in *Barns of Western Canada* (Fifth House, 1998): "A strange paradox of western barn hunting is that...there is not a 'Western Canada' barn form to be recognized apart from others....Just an array of shapes and forms adapted through ingenuity and individuality. The long, low barn signals a dairy operation, while an L-shape or U-shape frequently tells us success came to that farm in many stages, and wings were added as financial conditions allowed. An arched roof. The saltbox roof line. Large metal cupolas versus ornate wooden ventilation outlets. An extended roof line to accommodate a haysling track. Each gives a hint of the construction period. Other silhouettes such as connected house-barns talk plainly of a European heritage, most often in Mennonite settlements....Different ornaments or window arrangements tell viewers of other Old World origins."

It is hoped that this pictorial history of our irreplaceable rural heritage will reawaken a sense of the past in the present and foster new appreciation for the landmarks built with such skill and dedication by those who helped to shape the continent.

New England Farmsteads

The historic barns of New England have their origins in the building styles brought to the New World by early settlers from Europe. As these traditions were adapted and updated through the centuries, the classic barns of this region still reflect these early influences.

The earliest English farmsteads still extant may be seen in Chysauster, near Penzance, in the form of a row of two-room cottages built in the first century BC. One room sheltered the family and the other was a byre, or cowshed, with separate courtyards for each. This prototypical "house/barn" was built in various forms throughout Europe for centuries and imported to North America by ethnic groups including the Dutch, Germans and Swiss. However, in medieval England, the barn evolved into a building specialized for the threshing and storage of wheat, with livestock shelter—if any—as an ancillary function.

Since the British Isles were still heavily forested at that time, such barns were built of timber, along the lines described in the introduction. As timber grew scarcer, barns were constructed mainly of brick made of local clay, or from stone. However, they kept the same basic form: a central threshing floor flanked by large storage bays, with hinged doors facing one another on the broad side of the building to provide access to the threshing floor. Wagons laden with hay or grain could be offloaded and sent out through the opposite door without having to turn around. A steeply pitched gable roof covered the barn, which was often windowless. Many barns had a number of swallow holes cut high in the gable ends to encourage insect-eating swallows to nest in the building and keep it free of pests.

The basic three-bay English barn was the model used by colonial settlers of New England from the early seventeenth century. Originally clustered along the Massachusetts coastline and the Connecticut River Valley, they made several innovations dictated by the region's severe winter climate and the simplicity imposed by the absence of skilled master builders and timber framers. Colonial farmers combined the functions of the traditional grain barn and the cowshed, often using one bay for hay and grain storage and the other for livestock shelter. Oxen were the primary draft

animals at this time, and most families kept, or hoped to keep, a cow to provide milk, butter and cheese. Some farmers also kept a flock of sheep for the production of wool used in making clothing and blankets.

As in England, wheat straw or reeds were used to thatch both house and barn in early colonial times. Although these materials were not waterproof, they were bundled thickly and affixed to the rafters in layers to provide an insulating cover that shed rain and snow. This type of roofing was in use from French Canada to the English colonies of Virginia until it was replaced by split shingles, used also for siding. It is interesting to note that the traditional steeply pitched roofline persisted even in more southerly regions, where winter snowfall was not a factor.

Another introduction from England was the gambrel roof, with a double slope on each side, which provided more usable space in the loft area. Immigrants from other European countries adopted this form and took it across the continent as settlement extended westward after the Revolutionary War (1776–83). Thus the gambrel roof became closely identified with the North American barn. Much later, this type of roof was used on neo-colonial houses described as "Dutch Colonial," which created the mistaken impression that the double-sloped roof had originated in the Netherlands.

The typical New England barn was roughly twice as long as it was wide, often 60 by 30 feet. Vertical planks, clapboard, hewn logs and cedar shingles were all used to surface the walls, with planking predominant. Unfinished logs formed the floor joists, and the sills rested on fieldstone footings to prevent rot at ground level. Hardwood trees like white oak, ash and locust were preferred as building materials, but as these species became depleted, softwood timber, including spruce, pine and hemlock, were widely used. Even then, the treenails, or wooden pegs used for joining the timber frame, were cut from hardwood wherever possible.

New England farmers originally used the two side bays for storing threshed and unthreshed grain, with the hayloft located above the threshing floor for fodder storage. Supported on the barn's upper framing, the loft comprised a series of spaced poles that allowed for ventilation of the hay to prevent spoilage. Larger barns might have a two-bay threshing floor, over which the loft was supported by a powerful swing beam cut from a tree trunk up to sixty feet tall. The swing beam was the product of Yankee ingenuity—an innovation in vernacular architecture. It bore the weight of the haymow and also functioned as part of an X-shaped truss that held either hay or sheaves in place.

As animal husbandry became more important in rural New England, many early barns were modified to house cattle, dairy cows, sheep, or horses. One of the side bays might be given over to stalls, or an addition made to the original barn at right angles to form an L-shaped building. In some cases, several barns were moved end to end, as can be seen by the double main entrances on the eaves side in those that remain. Unfortunately, the number of old barns diminishes every year, as builders ransack the landscape for weathered boards and still-sturdy timbers from abandoned farms. Even barns still standing can be difficult to date, because farmers often salvaged old boards and beams to enlarge or alter their buildings over the years.

Valuable clues to the evolution of New England's rural architecture can be gleaned from field work like that undertaken by Thomas Durant Visser and his colleagues in preparing their *Field Guide to New England Barns and Farm Buildings* (University of New England Press, 1997). As they point out, "A careful examination of the recycled materials may reveal the type and size of the building that was torn down. Thus, a large barn built in the late nineteenth century might have hand-hewn beams from a farm's original late-eighteenth-century barn. Wear patterns on floor boards, empty mortises in timbers, and 'ghost lines' on walls may reveal missing interior features."

A typical late-eighteenth-century barn has large posts at the four corners and framing each bay, as well as inside posts that help support the structure by upholding the horizontal timbers of the haymow and the roof framing. The inside posts are often studded with rungs that served as ladders, especially the center post. Roof trusses may incorporate a single upright post in their triangular framing, in which case they are identified as king-post trusses. Those incorporating a pair of posts are referred to as queen-post trusses.

Many regional variations occur across the large area that eventually became the six New England states of

Previous pages: **Cobble Brook Farm, Kent, Connecticut;** *Above:* **Block Island, Rhode Island**

Massachusetts, Maine, Rhode Island, Connecticut, Vermont and New Hampshire. Since coastal eastern New England was settled first, the western area grew slowly in population between 1620 and 1790. Maine had especially close ties with French, and later British, Canada, and migrants moved freely between these areas. The state's boundary with Canada was not fixed by treaty until 1842. Most of Maine's agricultural land lies between the coastal lowlands and the White Mountains, and the variety of barns and outbuildings reflect the region's cross-cultural heritage.

Connected house/barns on the French model were adopted from Eastern Canada as appropriate to the region's severe winters: A farmer could lose his way — even his life — traversing his own farmyard during a blizzard. Annual snowfall in this area has reached 150 inches. Another type of continuous architecture, known as the New England barn, comprises a series of units grouped in line, or forming an L-shape. Common to the northeastern United States and Eastern Canada (as discussed more fully in chapter 4), it may include only a house, barn and woodshed, or take in a sheep-fold, dairy, poultry house, tool shed and other outbuildings sheltered by an irregular roofline. Most of these barns are of clapboard construction over timber frames formed by cedar or birch logs. An especially extensive complex of this kind is illustrated by Eric Sloane in his valuable reference *An Age of Barns* (Dodd, Mead & Co., 1985). It evolved between 1800 and 1850 in New Hampshire and eventually included shelters for sheep and swine, a springhouse, the main barn with adjacent dairy, a wagon and carriage shed, corncrib, henhouse, woodshed, the original single-story house and the two-story "new house" added in 1850.

Another style common to domestic and barn buildings of New England evolved in colonial Massachusetts, where houses were built with a long, low roofline facing north to shed snow and take the brunt of winter winds. Barns of this type were banked on the north side with insulation materials including sod, straw and cornhusks. Known as saltboxes because they resembled a medieval salt cellar, they were also built in Virginia and parts of the original Southeastern frontier. According to Eric Sloane, "This was an American idea, evolved to cope with American weather, and it seems to have been developed almost entirely by the pioneers who came from England. The very small saltbox barn appears often in the South; it is quite evident that the long sloped roof was not an afterthought [as in a lean-to addition], because the strongest wall was always the inner one … and the rafters were in one piece."

Many colonial barns had few or no windows, because the rough-sawn wide boards that covered them would shrink over time, creating gaps that lighted and ventilated the building. However, concerns about the health and comfort of livestock became more prevalent during the 1800s, and many farmers shingled their barns, or installed an additional layer of vertical sheathing in the form of thin boards tacked between the gaps in the exterior wall. Some favored double walls for stable areas, as recommended by one observer, who wrote:

Cold and open weather-boarded barns can easily be made warm by boarding them up on the inside and filling up space between the outside and inside weather boarding with straw or coarse refuse hay. And this can be done at a very trifling expense by such as cannot afford to build new barns or thoroughly repair their old ones. For a few dollars worth of boards and nails and a little work, which you can do yourself, is all that is necessary to prevent the ingress of the sharp winds and cold, frosty air. And he who neglects or begrudges this is unmerciful to his poor shivering beasts.

Originally, barn hinges were made of wood, as they had been in Europe. They consisted of a thick pin fitted into the door jamb and a strap with a hole on the end that fitted over the pin. Americans called this a "pintail" hinge, later shortened to "pintle." When blacksmiths and nailors began to produce barn hardware in wrought iron, they used this design as a model. Later, more elaborate and ornate hardware in the form of door hinges, bolts, nails and other fastenings were produced by smiths, or by farmers who bought iron rods and straps from itinerant peddlers and ironmongers. Especially large nails were needed to fasten the threshing-floor (later, drive-floor) doors that took hard use: Their boards tended to loosen over time. Flat-sided "cut" nails were made an inch or more longer than the door was thick. When hammered through, the sharp points protruded inside the barn; the carpenter flattened them in a process called clenching, which secured the boards closer together. A nail clenched in this manner was called "dead," since it could not be reused, and we gained the vernacular phrase "as dead as a doornail."

Originally, wooden barns were left to weather naturally, but as cedar and spruce clapboards became more popular—no longer hand-riven, but cut by a gang-saw—more barns were painted with a homemade mixture of red iron oxide, skim milk, lime and linseed oil. Later, inexpensive commercial paints of this kind became available. Long before the Civil War, the custom of painting barns red with a contrasting white trim was well entrenched, increasing their durability and resistance to moisture as well as their attractiveness.

One type of New England barn was never painted and remained largely unfinished, including timbers from which the bark had not been removed. This was the tobacco barn of the Connecticut River Valley, where shade-grown broadleaf tobacco became a profitable crop during the seventeenth century. By the early 1800s, New England tobacco was in demand as the premier cigar wrapper, and many farmers turned to this risky but well-paid form of agriculture. Cultivated under acres of netting strung on poles, the tobacco leaves had to be dried slowly after harvesting so as not to become too brittle. The barns built for this purpose averaged 30 by 100 feet, rising some 20 feet to the eaves. The bents, or frame sections, were spaced at 15-foot intervals. Four tiers of cross braces were installed the full width of the barn, and the poles on which the tobacco was hung ran the length of the building.

Tobacco barns were ventilated by hinging alternate boards of vertical siding near the top, which could be swung out several feet as needed. Later versions had horizontal siding that could be opened in sections via a vertical tie-piece. Access was through wide hinged doors at the gable ends, and some buildings had roof-mounted ventilators. In addition to the hazards of the curing

Philipsburg Manor, North Tarrytown, New York

Bradley-Wheeler Barn in Westport, Connecticut

process, the tobacco farmer also had to contend with plant diseases, tobacco worms and a boom-and-bust cycle for his product. In 1874 the influential *American Agriculturist* carried this moralizing opinion: "The magnificent castles in the air which have been erected during the past few years by the over-sanguine tobacco growers now lie in ruins. The unfortunate builders are disappointed and disgusted. It was ever thus with growers of specialty crops. For a few years large profits tempt greater ventures and then come excessive crops for one or two seasons and the prices go out of sight." Another commentator deplored the fact that "The acres that ought to be growing fruit and vegetables for human sustenance are grazing the tobacco worm, and farmers are laboring to check its depredations."

Before the tobacco business in the Connecticut River Valley was largely overtaken by new mechanized methods of wrapping and binding cigars, many impressive barns were built, some of which are still standing. They include the large slate-roofed barn in South Windsor, Connecticut; the substantial working barn in Whately, Massachusetts, where propane gas is now used to control the temperature for air-curing; and the combined tobacco stripping and curing facility at East Windsor, Connecticut, built on a raised, brick-walled foundation used as a cellar in which to keep the leaves soft for easy stripping before they are dried.

According to Nicholas Howe, the author of *Barns*, the historic community of Deerfield, Massachusetts, had no fewer than four working farms well into the twentieth century. Founded in 1669, the small village was a frontier outpost in British North America. It was nearly destroyed by a French and Indian war party in 1704, but was soon rebuilt and still stands as a living replica of colonial and post-Revolutionary New England. Its Wells Farm has a very large main barn dating to the late 1700s. At the Cowles Farm, the lower level is lined with passages behind the animal pens for ease of feeding and cleaning, and the main floor is large and airy for hay storage. It also contains a bull pen.

On a ridge above town, a Dutch immigrant named Frick Spruyt settled with his American wife in the early twentieth century and maintained a house/barn connected by a toolshed. Mr. Howe, who grew up in Deerfield, describes the Spruyt barn as "a vast place of cool darkness lit only by thin shafts of light coming in through the chinks; there was a haymow several stories high, and each kind of animal had its own section of the barn. The goats, for instance, had a long, narrow wing with many small windows and immaculate pine-paneled stalls, each topped by a brass fixture that held a slip of paper printed with the name of the resident." The Spruyts told the old legend that on Christmas Eve at midnight, the animals could speak in honor of the Christ Child, but "the gift would come only if the right carols were sung, a different one for each kind of animal. After dinner we'd all walk out into the snow and the night, and go to the barn and sing to the goats and sheep and horses and oxen….Then we'd leave quickly, because if any humans were around to hear, the gift of talking would not come." Another version of this legend says that all the animals kneel at midnight on Christmas Eve, recalling their presence in the stable at Bethlehem on the holy night.

The nineteenth century brought many changes in agriculture across the continent. In New England, some farmers turned to dairy farming where the land was too hilly or rocky to grow crops profitably. Vermont, in particular, specialized in dairy herds; it became known as the state with more cows than people. The barn was redesigned or rebuilt to facilitate the milking process and the storage of bulk quantities of milk for transport to urban centers. Ventilation and sanitation were improved with the introduction of air shafts terminating in louvered rooftop cupolas, some of which also served as chutes for delivery of fodder from the hayloft to the ground-floor stalls. Most of these farms made their own butter and cheese in a milk room, or creamery, usually located off the kitchen and sometimes banked into a hillside for coolness.

In the dairy, several tiers of shelves lined the walls to hold shallow pans in which the milk stood until the cream rose to the top. It was then skimmed off and churned into butter. During the late 1800s, many dairy farmers added ventilators to these rooms to maintain even temperatures and installed sinks to wash their equipment, fed by gravity flow from a nearby water source. The walls were whitewashed with lime to keep the building clean; some were plastered, or furnished with glazed tile, in the manner of the best English dairies, of which the pre-eminent example was the Royal Dairy at Windsor, designed by John Thomas for Queen Victoria and Prince Albert. This delightful building is now in the process of restoration by Great Britain's National Trust. Yankee farmers might have been astonished at its ornate tile mosaics, china and plaster oranges entwined with colored ribbons in relief, twin fountains and stained-glass windows, but they could appreciate the sound practical features introduced by Prince Albert. These included double-glazed windows to maintain even temperatures and a network of pipes beneath the wide tile floors that cooled the building.

A separate room for drying cheeses often formed part of the dairy farm complex. As early as 1797, Samuel Deane, author of *The Newengland Farmer*, recommended:

As to drying cheeses, [they] should never be kept to dry in the same room where milk is set, for they will undoubtedly communicate an acidity to the surrounding air, which will tend to turn all the milk sour….And a drier room would be better for cheeses; only let it be kept dark, that the flies may not come at them.

During the 1880s, the introduction of the centrifugal cream separator increased the productivity of the regional dairy farm. Shortly before this, the horse-powered churn had facilitated the labor-intensive process of making butter. Most farmers had developed their herds to thirty or more cows by this time, and co-operative creameries were being established. According to the *Field Guide to New England Barns and Farm Buildings*, such creameries were "usually located next to the railroad line in villages. [They] processed the milk of dozens of farmers, who shipped the liquid from the farm to the creamery by wagon in metal milk cans." In 1887 the journal *New England Farmer* praised this development: "The fact that co-operative dairies have now been in operation for over five years, and as yet there has been no failure, is the highest possible recommendation of the system and speaks volumes for the business capacity of the dairymen of New England."

Many field and hay barns also sprang up on regional farmsteads during this period. For generations, hay had been New England's principal crop, and there were several advantages to constructing the simple hay storage barns at some distance from the main barn. The *American Agriculturist* observed in 1842 that "Some farmers…build small hay barns which stand scattered about the fields contiguous to springs or water courses, for storing hay as it is cut, from which it is fed out in winter, and the cattle driven to them for that purpose." Another advantage was the ease of transporting loads of hay to the main barn by sled once snow covered the fields. Thus the fodder consumed by the herds could be replenished throughout the winter months. As many farmers found their old English barns inadequate, some of them were moved to serve as hay barns; other field barns were used to shelter sheep, especially during the wool-market boom of the early nineteenth century.

As affluence increased, many merchants and wealthy gentleman farmers constructed "model farms" that reflected the prevailing influence of Victorian architecture. During the 1850s, a New York City businessman named David Leavitt built an imposing complex at Great Barrington, Massachusetts. Its main barn was in the Gothic Revival style and measured 200 feet long. It spanned a ravine, from which an embankment gave access to the topmost level of the three-story building. Constructed at a cost of $20,000, Leavitt's Cascade Barn overlooked the Housatonic River and was pictured in the 1855 book *Agriculture in Massachusetts*.

The Morgan Horse Farm at Weybridge, Vermont, had an elegant French Second Empire stable built in 1878 as the centerpiece of an enterprise devoted to the breeding and revival of a popular American horse descended from a single stallion named Justin Morgan. Foaled in the late eighteenth century, he belonged to a Vermont schoolteacher of the same name and is believed to have been sired by a son of the thoroughbred Morton's Traveler and a mare of thoroughbred descent. According to *Modern Breeds of Livestock*, by Hilton M. and Dinus M. Briggs (Macmillan, 1969), "Justin Morgan distinguished himself for his ability to perform, and it was said that he was able to outrun, outpull, outwalk and out-trot horses with whom he was matched in Vermont. His outstanding ability seems to have been to win races of a length not exceeding a quarter of a mile....Remarkably sound and energetic, the little horse lived to be thirty-two years of age."

The riding and driving ability of his offspring made the Morgan horse one of the most important light horses of the day, but the breed declined due to the lack of mares of the Justin Morgan type. In 1870 the Middlebury, Vermont, breeder Colonel Battell began his lifelong effort to trace the descendants of Justin Morgan and rejuvenate this remarkable breed, enlisting the cooperation of the U.S. Department of Agriculture, to which he deeded over the Morgan Horse Farm in 1907. Thirty-five years later, the farm was turned over to the University of Vermont.

The nation's grandest barn is the massive Farm Barn at Shelburne, Vermont, studded with turrets, half-timbered gables painted green and conical tower roofs in the Norman Revival style of the late nineteenth century. The Shelburne Farms complex was founded during the 1880s by Dr. W. Seward Webb and his wife Eliza Vanderbilt Webb, both people of taste and vision as well as enormous fortunes. They purchased adjacent farms to increase the original property to some 3,600 square miles of the Lake Champlain area and engaged landscape architect Frederick Law Olmstead to create a design for fields and forests. New York City architect Robert Henderson Robertson drew up the original plans for Shelburne House as well as the farm's palatial barns and outbuildings.

Devoted to the ideals of "scientific agriculture," the Webbs built the enormous Breeding Barn in which to develop a new breed of horse, combining the traits of the English hackney and the Vermont draft horse, which could be used as both a field and a carriage horse. This expensive project, housed in a building so large that it doubled as an indoor polo field, proved impossible to implement, but the now-weathered gray barn still stands. Like the adjacent Coach and Farm Barns, it is now being restored under the auspices of the nonprofit Shelburne Farms organization, which acquired the estate from the Webbs' heirs, who wished to see it preserved. It is a working farm again, complete with a herd of Brown Swiss dairy cows and other livestock, and children visit on field trips every day of the school year to participate in its environmental and educational programs.

New England's rural landscape has long been defined by its fieldstone walls, laid up without mortar in the masonry form known as drywall. They circled its towns, divided its pastures and lined its woodland roads. The region's stony soil is the legacy of glaciers that retreated some 12,000 years ago, leaving chunks of granite, gneiss and schist that are still heaved up by deep winter frosts. In colonial times, these stones had to be cleared from the fields laboriously before cultivation could begin; some local farmers still refer to them ruefully as "New England potatoes."

Many of the stones were incorporated into barn footings and foundations, or used to build springhouses and other outbuildings, especially as the region became deforested. By the mid-nineteenth century, three-quarters of the land had been cleared for crops and livestock, and the stone wall was well entrenched as an economical boundary marker. Its natural beauty is still a picturesque feature of the landscape, celebrated by poets and artists, while scientists have identified it as the repository of a unique ecosystem that attracts and supports many kinds of wildlife, from songbirds to snakes and short-tailed shrews. It is estimated that some quarter of a million miles of stone walls once crisscrossed New England and adjacent New York State. Unfortunately, many of these rural landmarks are rapidly disappearing from our landscapes. Many are now being dismantled, like the old farmsteads, for conversion of the agricultural landscape to the development of new suburban housing tracts.

The lighter side of New England farming has been preserved in a delightful book entitled *Life in Danbury*, by a Connecticut journalist named James M. Bailey, who called himself "the Danbury News Man." Published in Boston by Shepard and Gill in 1873, the book is a first-hand humorous account of the trials of local farmers and townsfolk:

> *The drive to Waramaug Lake from New Milford is through a pleasant if not a profitable country. Man doesn't appear to have extraordinary good luck with the soil, but nature is getting rich from it. Her trees and smart weed and Canada thistle are doing as finely as any I have yet seen.*

Further on, he complains that: "The meat markets are ill supplied with palatable stock. There is but little veal, and the beef is tougher than losing a mother."

His advice on livestock handling includes the following amusing suggestion: "A Brookfield man writes for the best way to manage a bull. If our Brookfield friend has got a bull on his premises, and the bull is well, he don't want to manage it. All he has got to do is to get a few things hastily together, mortgage his place and steer straight for the West. He might as well try to ward off a stroke of lightning with a fifty-cent paint brush as to manage a bull."

In fact, by this time, many New England farmers were pulling up stakes to try their luck in the burgeoning West, and their migrations can be traced to farmsteads as far away as the fertile valleys of Washington State.

Bethlehem, Connecticut

Connected Barns *Opposite*

New England's winters—severe by comparison with Great Britain's mild winter season—gave rise to farmsteads whose shelters were closely connected, so that the farmer and his family could do their many essential chores without exposing themselves and their livestock to the frigid temperatures, ice, bitter winds and snow.

Traditional Three-Bay Barn *Below*

Low-lying Block Island, Rhode Island, is the site of this handsome traditional three-bay barn, which is raised on supports to avert the danger of flooding during storms. The three front-facing gables and the full-size windows are unusual features, probably added during twentieth-century renovations.

Maple Sugaring in the Berkshires *Opposite*
Late winter at the South Face Farm in Ashfield,
Massachusetts, is the time when the rising sap in the
maple trees is collected in buckets and boiled in large
evaporator pans for many hours. This operation was
performed outdoors in iron kettles until the mid-
1800s, when single-story sugar houses came into use.

Winter Pastoral *Below*
A neatly fenced Massachusetts farmyard is
dominated by a well-maintained barn whose red
siding stands out against the snow. The custom of
painting barns red originated in Sweden and
eventually became the norm in North America.

New England Banked Barn *Below*
During the nineteenth century, many Northeastern farmers adopted the Pennsylvania-style banked barn, which is built into a hillside for improved weather protection and is usually three stories high. This handsome example in Waitsfield, Vermont, has an unusual dormer window in the attic-level loft and a distinctive cupola.

The L-shaped Barn *Above*

As New England farmers prospered, many built
an extension at right angles to the original barn,
providing more space and a farmyard partially
sheltered from the wind. Here, an earthen ramp
leads up to the main door, which is at the gable end.

Side by Side *Overleaf*

In picturesque Lyndonville, Vermont, roomy twin
barns are reflected in a farm pond—a tranquil scene
that belies the rigors involved in maintaining a
farmstead to this standard.

Washington Depot, Connecticut *Above*
Many Pennsylvania-style barns were constructed throughout the New England states during the nineteenth century. This L-shaped example on a fieldstone foundation has a pentroof supported on sturdy wooden posts. Partially enclosed, it shelters animals stabled on the ground level.

A Massachusetts Sugar House *Opposite*
Steam from the boiling sap is vented by a flue in this typical outbuilding with gabled roof and vertical siding. The sugar house was always separate from the main barn because of the danger of fire. Often, it was located in the maple sugar grove (called the "sugarbush") for convenience.

The Country Place *Below*

This handsome nineteenth-century barn in historic Stockbridge, Massachusetts, reflects the town's popularity as "the inland Newport" with the wealthy summer colony that came from New York and Boston to enjoy rural life in comfort amid the scenic Berkshires. In 1871 *Harper's New Monthly Magazine* described Stockbridge as New England's "most beautifully set jewel."

Extended Barns *Above*

Additions on either side of the original three-bay English barn at Casey Farm (c. 1840), in Saunderstown, Rhode Island, mark its growth and progress as the farm prospered and expanded.

Essence of New England *Overleaf*

A classic red gambrel-roofed "double decker" barn and outbuildings flank the historic Congregational Church at Peacham, Vermont, in this quintessential New England view.

Part of the Landscape *Below*
Weathered vertical planking and a louvered
cupola partly obscured by trees mark a Cornwall,
Connecticut, barn being rapidly overtaken by
disuse in modern times.

Pittstown, New York *Above*
The influence of the New England style is widely apparent throughout rural areas of adjacent New York State, as seen in this updated working building used mainly as a hay barn.

Fall's Colors *Opposite*
A venerable farmstead near Cavendish, Vermont, is framed by blazing maple trees. Vermont is the largest producer of maple syrup in the United States and the nation's most rural state, with around two-thirds of its population living in towns with 2,500 residents or fewer.

The House/Barn Influence *Overleaf*
A multifamily dwelling in Lancaster, New Hampshire, is backed by a gambrel-roofed dairy barn with massive steel ventilators that were introduced after the turn of the century.

Ornamental Doors and Siding *Below*
A split-rail fence defines the farmyard of this sensitively modernized barn in Peru, Vermont, which is distinguished by its bold Dutch doors and diamond-patterned sidewall.

Dutch-style and "Pennsylvania" Barns

Previous pages: Amish barn in Intercourse, Pennsylvania; *Above:* Banked barn, Morristown, New Jersey

The most sophisticated early colonial barns were those built by Dutch settlers during the seventeenth and eighteenth centuries, beginning in the Hudson, Mohawk and Schoharie River Valleys of present-day New York State. During medieval times, the northwestern part of the Netherlands, or Holland, was heavily forested: In fact, the name Holland is probably derived from *Houtland*, meaning "Wooded Land." The availability of timber for building, combined with a soft subsoil that discouraged the use of heavy masonry materials, made carpentry a master art.

Great churches, halls and barns were built for both the nobility and the monastic orders on the basilica plan of early Christian churches, that is, a wide central nave flanked by aisles. The monastic "tithe barns" were used to store the portion of grain contributed by every landholder to the Church by law. In the Dutch method of timber framing, no tie-beams were used to support the lofty roofs; heavy wall posts carried most of the lateral thrust. Entrance to the wide threshing floor of the barn was from the gable end—a feature unique to Dutch barns built later in North America, most others having their entrances on the wide, or eaves, side. The aisles were used to stable horses, cows and other

livestock, and the haymow was located above the threshing floor. The animals faced the center of the building, from which they were fed, and side doors provided additional access to the stalls and "nave."

The earliest Dutch farmsteads, like those in nearby Saxony, were rectangular and had steep roofs that were hipped at either end. They sheltered both people and animals, combining barn, byre (cow barn) and house. Thatched roofs shed snow and kept the interior relatively warm and dry. A slow-burning peat fire smoldered on a low hearth at the end of the threshing floor for heat and cooking but it provided the only light by night. No lamps or candles were used, for fear of fire, as in medieval England, where country people lived by the proverb, "A lantern on the table is death in the stable."

Sleeping quarters for the family and servants were above the stalls, and windows in the living area were small. Before glazing was introduced, windows were covered with translucent cured sheepskins. In the absence of chimneys, a small hole in the roof vented smoke—or did not—depending on the wind's direction. Over time, the discomfort of these murky, close interiors was somewhat alleviated by the addition of ground-floor living rooms with cupboard-style beds built onto the end of the structure. Examples of these prototypical house/barns (which occurred in several forms throughout Europe from the Iron Age onward) may be seen at the open-air museum in Arnhem, where the Netherlands government had them moved and restored faithfully in every detail. According to Eric Arthur, the author of *The Barn: A Vanishing Landmark in North America* (Arrowood Press, 1989):

"Among the oldest [is] the one from Lichtenvoorde. It is basilican in plan, and one can look from the wagon entrance through the threshing floor to the ever-burning fire. On entering, two rooms for pigs are passed on the right and a sunken room for cattle on the left. Research has not unearthed other barn types in Europe or elsewhere where the cattle stall floors were depressed, but the custom born of centuries was continued in New York State....The cobbled floor of the family area surrounding the hearth stops abruptly on a line beyond which is the earth-packed smooth threshing floor. Over all was the pole ceiling supporting the hay."

Dutch timber trade and construction were fostered by increasing use of the windmill, as discussed more fully in chapter 7, after 1600. In Zeeland, Brabant and Flanders, the basic "aisle barn" was clad in tarred boards for weatherproofing, and the door surrounds were whitewashed. Sawmilled lumber was also used in North Holland, where farmsteads comprised a large hay shed under a low-eaved pyramidal roof, around which people and animals were housed. Wooden cladding helped to preserve the fodder essential to the livestock's survival through the winter months.

Dutch settlers in the New World made several innovations based on changing needs, climate and available materials. Eventually, their distinctive gable-end barns would be seen from New Amsterdam, now New York City, to Ontario. New Jersey, too, still has its share of rugged Dutch barns dating back to pre-Revolutionary times. A major change was the abandonment of the *Loshoes*, or house/barn plan, in favor of separate buildings. With living quarters removed, it became possible to build a drive-floor exit in the gable opposite the entrance, so that incoming wagons need not turn around on the threshing floor. The new entry also provided better air circulation for the process of winnowing the grain from the chaff by hand.

In the Old World, the walls had comprised timber posts infilled with brick, or by the ancient wattle-and-daub method: woven sticks daubed with mud or clay, and later, plaster. In North America, wide clapboards were used over the timber studs. The eaves were raised,

Fieldstone hillside barn, Leacock, Pennsylvania

Blizzard conditions in Roosevelt, New Jersey

and the former rectangular plan was replaced by one that was nearly square. An example is the Somers barn in Ontario's historic Upper Canada Village, which measures 50 by 45 feet. Roofs of reed thatching were replaced by cedar shingling, and the hipped roof was abandoned in favor of gables. Earthen threshing floors were covered with thick planking, and side doors in the gable ends allowed for the animals to enter the stalls. These two-part Dutch doors also provided better ventilation, as the top could be left open without allowing the stock to wander in and out at will.

Massive anchor beams up to 30 feet long supported the hayloft, dividing the Dutch barn into bays parallel to the gable ends. The roofline was less steeply pitched than that of the English barn, and had only two slopes, but the loft could store a great quantity of hay. This was loaded from ground level and piled loose on movable poles laid across the anchor beams—sometimes the same spike-tipped poles that had been used to help raise the bents during construction.

Finished timbers were the norm for Dutch barns in North America, whereas trees crudely stripped of their branches and left unsawn were often used in the Netherlands. This custom originated with the ancient "cruck barn," in which two sets of bent tree trunks were used to form a cruck, or juncture, at each end of the building and braced by a ridgepole that

formed the apex of the roof. Stone footings supported the crucks, and horizontal tie and collar beams helped carry the roof, which sloped down from the ridgepole on either side. Occasionally, Dutch settlers in the New World used a gambrel roof with distinctively flared eaves. The main doors often had a smaller door set into them for easy access to the barn, especially in cold weather. Until the advent of roller doors during the 1840s, most American barns had hinged doors or sliding panels, and opening the main doors was a cumbersome business. The earliest examples had a large wooden partition that was propped shut with a long beam all winter long and removed for the summer months.

Early Dutch immigrants brought the ancestors of the Holstein-Friesian dairy cattle that would play such a prominent part in the American dairy industry. This popular breed originated in the Netherlands provinces of West Friesland and North Holland, where it was bred for optimum milk production, large size, robust health and sturdy offspring. Eventually, these cattle were exported to northern France and eastward into the German province of Schleswig-Holstein. Usually black and white, sometimes red and white, the Holstein, as the breed was called in the United States, consistently produced record quantities of milk with the desirable high quantity of butterfat.

The breed's most important foundation sire—though by no means the earliest—was the bull Netherland Prince, imported by the Lakeside Farm of Syracuse, New York, in the 1880s. At this time, the Holstein was introduced to Canada, where it soon surpassed the popularity of the English-bred Jersey. Several notable estate farms contributed to the breed, including Winterthur Farm in Delaware, established by Colonel H.A. du Pont. According to Clive Aslet, author of *The American Country House* (Yale University Press, 1990): "In 1914, when Henry Francis du Pont inherited the family estate...the 2,000-acre farm included a pedigree herd of Holstein-Friesian dairy cows, some 250 Herefords [beef cattle], a hundred prize hogs, an equal number of Dorset sheep and more than 2,000 poultry. Within four years, du Pont had rebuilt his dairy barns with long, low hipped and gabled roofs and a great central cupola, all providing for complete reventilation of the barns every six minutes."

This concern with ventilation and sanitation reflected the many improvements in American barns that had taken place during the nineteenth century. Agricultural publications including *The American Agriculturist* and such books as *Barn Plans and Outbuildings*, published by Orange Judd Company in 1881, kept American farmers apprised of new developments in animal husbandry, crop production, machinery, storage and marketing. And the late-century trend toward the construction of large country estates with model farms, like Winterthur, had a far-reaching effect on agriculture in general.

At Biltmore, North Carolina, for example, George W. Vanderbilt retained the eminent architect Richard Morris Hunt and landscape architect Calvert Vaux to construct a country house that combined the features of the French Renaissance chateau of the Loire Valley region and the great landscape parks of the Victorian-era English country estate. Described in 1908 as "probably the largest and finest estate in America," Biltmore comprised 125,000 acres, and the façade of the main house was 374 feet long. Subsidiary buildings included lodges, stables and a dairy, for all of which the stone was quarried nearby. Edith Stuyvesant Vanderbilt, the lady of the manor, acted as a patroness

of local agriculture, which prior to the Vanderbilts' arrival had consisted mainly of unprosperous small holdings from which the mountain people scratched a meager living, depleting the soil in the process. However, despite initial resistance from the local country people, many improvements took place in this region's farming methods. Two years after the estate was completed (1888–95), the *Asheville News and Hotel Reporter* observed that "It is Vanderbilt the farmer, not Vanderbilt of the Chateau, who has proven to be the great benefactor of Western North Carolina. He has shown the Carolinians the productive capacities of the virgin soil...by the scientific drainage, the improved machinery, the importation of fine stock, the judicious and lavish use of fertilizers, and the most up-to-date and scientific methods of farming."

To explore the origins of the better-known "Pennsylvania barn," we return to an earlier period, when German immigrants arrived in the Delaware Valley to build "banked" or hillside barns similar to those in their European homelands. Eventually, the Pennsylvania-style barn would be constructed as far south as the southern Appalachians and as far north as western Ontario. At the outset, however, it was the closely related German and Swiss barns that provided the model for the shelters first built in colonial Pennsylvania. In the mountainous regions of Upper Bavaria, the Black Forest and Switzerland's Jura region, it was customary to build the barn into a hillside, with entrances at several levels, the main doors being accessed by a ramp. Heavy timbers formed the framework of these two-story barns, in which the livestock was housed at ground level, with the threshing floor and hayloft above. Stone was often used as walling, while a log forebay jutted from the second level, which usually faced south. Thus it sheltered the farmyard and the animals within from prevailing winds during the winter months and provided shade through the summer. Many of these structures were, in fact, long house/barns built at right angles to the slope, with the two-story house sharing a common roof at the gable end, but comprising a separate structure that was often elaborately carved and colorfully painted.

German and Swiss settlers in the New World made several important innovations, the first of which was to separate the dwelling from the barn entirely, as the Dutch had done. In fact, as elsewhere, the barn was often a far more substantial building than the original house, since the family's livelihood depended upon the safety of its stored crops and livestock. Another important change was construction of barns that were set parallel to the hillside rather than at right angles to it. In these banked buildings, the drive doors were either level with the hilltop, or reached by a wide earthen ramp. Fieldstone was often the primary building material. Where clay was available, brick was made to serve as nogging, or infill, between the structural members of timber. During the eighteenth century, itinerant masons and bricklayers traveled from farm to farm to help with the masonry components of the building. Sometimes their names, or the date of construction, are carved in a door lintel, just as the master carpenter left his mark on the wooden armature with an adze or a chisel. Stone barns were less susceptible to fire than those built entirely of log or frame, and at this time such a fire was still the farmer's greatest fear. In fact, colonial law forbade setting a fire within thirty feet of a barn.

The blacksmith's forge was always in a separate outbuilding, usually constructed of brick or stone. The smith not only shod the horses, but produced a wide variety of hand-crafted hardware and the tools and machinery needed for building and agriculture. Few farms were large enough to support their own forge. Most used itinerant blacksmiths, or ironworkers located in nearby villages who served a wide area.

Like the Dutch barn, the Pennsylvania barn was basilican, with aisles that flanked the threshing floor, but the main entrances were on the wide side of the building. The second-story galleries were used for hay and grain storage, which extended into the forebay, while the ground level had stalls for horses, oxen, milk cows and other livestock. Root cellars for winter storage of vegetables, cider and "keeping" fruits like apples were located at ground level, or dug into an adjacent hillside as separate facilities. The overhanging forebay was supported on timber joists that extended from the main floor, much as Spanish roof beams, called *vigas*, protrude from the façade of adobe buildings of the Southwest. Where timber of sufficient strength to support the forebay was lacking, an overhanging pentroof that projected several feet over the farmyard was used instead. Similar pentroofs protected the smaller entrances from rain and snow.

Early Pennsylvania barns dating from the 1700s, of which few examples remain, usually had second stories of log construction. Over time, gable ends or entire buildings of stone or brick became the norm, and lighter frame construction replaced notched timbers. A rare example of the early log banked barn has been preserved at Black Creek Pioneer Village, near Toronto, Ontario. Many agricultural and folk museums of this kind have helped to rescue endangered landmarks of vernacular architecture for a generation of young people who may never have seen a working farm. In Berks County, Pennsylvania, the rebuilt Village Barn at the Hopewell Village National Historic Site provides a firsthand experience of the typical Pennsylvania bank barn of fieldstone construction, with its forebay, ramp on the uphill side and stables at ground level. Louvered openings at both levels provide for ventilation. Additional forebay support was often provided by columns of stone, brick or timber, or by an end wall that helped shelter the farmyard when the animals were turned out, or workers were hitching their oxen or mules to wagons or plowing equipment.

Another historic German-style barn, dating to 1788, is the one still standing at the Dundore Farm near Mount Pleasant, Pennsylvania. This frame and stone building has an unusual capacity for this early date, measuring some 125 feet long by 40 feet wide (including forebay). The front elevation is almost entirely banked up to provide access to the hay storage areas through a series of drive doors with smaller doors inset. This unusual barn has no fewer than six bays that were used for storage and threshing. Some fifty years later, a separate granary was constructed, with tools and wagons stored on the first floor and the valuable grain in second-story bins secured against rodents and other animals. The granary, too, is banked into a hillside and constructed of wood and random limestone. Such limestone was also burned in kilns as a primary ingredient for mortar for bricklaying.

An early nineteenth-century barn at Shawnee in Delaware, Pennsylvania, was built into a hillside on the Walter-Kautz Farm. A frame structure on a stone foundation, it has three levels, with sliding doors—a recent introduction at the time—on the front ramp entrance. A shed addition at the rear forms an end wall for a forebay supported by pillars, and a later addition probably served as a henhouse or piggery. A side view

of the structure, on which much of the siding has deteriorated, reveals that the basement level was used for small animals and root storage; the main floor for cows, horses and threshing; and the loft for storage of hay and straw. By the close of the nineteenth century, mortise-and-tenon timber framing would be largely replaced by the new and more rapid method of balloon framing, and the use of rooftop cupolas would increase both light and ventilation inside the barn. The former narrow ventilation slits built into masonry barns, and the picturesque patterns formed by omitting bricks to admit air and light, gave way to small glazed windows and transom lights over the main doors, which had formerly been covered by a movable board.

Meanwhile, banked barns had been widely adopted by New England farmers and their neighbors in eastern Canada, and families of German origin took their traditional barn designs south and west as the frontier expanded from the Appalachians to the Mississippi River and beyond. Additional examples of their sturdy and efficient farmsteads appear in the following chapter, which includes some of the Utopian communes established by the Mennonite, Amish and Hutterite sects, whose impact on North American agriculture was out of all proportion to their numbers.

Opposite: **Fieldstone wall detail, Kutztown, Pennsylvania;** *Above:* **Large Dutch-style barn, Rheems, Pennsylvania**

Germanic Heritage *Opposite*

An Amish buggy drawn by a Standardbred horse
passes a whitewashed banked barn with vertical
siding and a sheltering forebay in New Holland,
Pennsylvania. The original roofing has been
replaced by metal in modern times.

Old-fashioned Feed Storage *Above*

The free-standing fodder-storage facilities on this
Mackeville, Pennsylvania, farm, as well as the hay
barn, show years of weathering. The left-hand tower
of the group of three corn cribs has tilted on its
foundation, while the advertisement painted on the
hay barn is now quaintly outdated.

Hinged Panels for Ventilation *Overleaf*

This unusual three-bay barn in Bethel,
Pennsylvania, flanked by what appears to have been
a poultry house, resembles a tobacco barn in its use
of hinged siding for ventilation.

An Historic Dutch Mill *Above*
Solid as a fortress, this five-story mill of fieldstone construction, with various silos added on, is a landmark in Lititz, Pennsylvania.

Gable-end Stone Barn *Opposite*
Dutch and German features combine in this three-story barn with a pulley for lifting hay to the top level. Beyond the grazing calves, a pentroof extending along one side of the barn is visible.

Bow-truss Roofline *Below*

The semicircular bow-truss roof, introduced in the early 1900s, became popular because it provided a larger-than-ever hay storage capacity and was easily assembled. The bow-truss roof is an adaptation of the flared eaves introduced to North America by Dutch settlers. This example is in Barron County, Wisconsin.

A Dutch-style Dairy Farm *Above*
Triple metal ventilators line the roof of an early
nineteenth-century dairy barn in Wisconsin, with
stabling on the ground level augmented by a large
cow shed added at a later date.

Amish Country *Overleaf*
An unusually large dairy farm houses several
generations of a Somerset, Pennsylvania, family
and their well-kept herd of Holstein-Friesians.
The handsome symmetrical banked barn has
paired louvers for ventilation at the upper level.

A Study in Roof Styles *Right*
This spacious Midwestern connected barn includes a
central gambrel-roofed portion with a lean-to wagon
shed and gable-roofed addition. A large gabled wing
and a long single-story shed with fenced farmyard
project on the left side.

**Banked Barn, "Pennsylvania Dutch"
Decoration** *Below*
A long forebay overhangs and shelters the
stableyard of this German-style barn in New
Smithville, Pennsylvania, painted with traditional
"hex signs" in the popular star motif.

Additions and Outbuildings *Overleaf*
A trim red-and-white toolshed, dairy and other
outbuildings cluster around a gambrel-roofed barn
with windowed lean-to additions. The old-fashioned
sleigh and wagon date from the nineteenth century.

In the Gothic Mode *Right*
This delightful banked barn in Highspire,
Pennsylvania, has an unusual front-facing gable
adorned with a louvered star and pointed louvers
on the forebay and cupola. The "Rustic Pointed" or
Gothic style was popularized by Alexander Jackson
Downing during the mid-1800s.

Mural Artwork *Below*
The Wayne L. Weisner barn (1888) in Krumsville,
Pennsylvania, has colorful murals of the farm's dairy
cows and mules along with traditional hex signs,
a pair of American flags and a sliding hay door
in Christmas dress.

A Growing Concern *Overleaf*

This Amish farmstead in Lancaster, Pennsylvania, has added space—silos and housing—since the nineteenth century, as its founding family increased in numbers and prosperity. Note the absence of electrical power lines: The Old Order Amish do not use either electricity or internal-combustion engines.

Livestock and Feed Storage *Above*
This thriving complex at White Horse,
Pennsylvania, houses its livestock in the long
extension to the left of the main gable-roofed barn
and has three silos and a conical-roofed corn crib
to store green and dried fodder.

Farm Pond Reflections *Overleaf*
A simple L-shaped storage barn on a stone foundation and a gable-roofed shed are mirrored in an adjacent pond at Hamburg, Pennsylvania.

At the Stable Doors *Below*
An Amish farmer in Intercourse, Pennsylvania, relies on his horses for draft and road work, as most American farmers did well into the twentieth century. These sturdy workhorses relax outside their·stable after a hard day.

The First Frontier

As settlement spread west and south of the original thirteen colonies, new forms of vernacular architecture drawn from central European and Scandinavian models appeared in the Delaware Valley region and, eventually, the Deep South. Here, the log cabin and barn so closely identified with later frontier life were built as early as 1638 east of the Appalachian Mountains. Germans who had originally settled in William Penn's colony of Pennsylvania migrated south to present-day Delaware, Maryland and adjacent Tidewater states. The Swedes settled first in the Delaware Valley, and the two culture groups borrowed freely from one another and from the English settlers around them.

According to Richard Rawson, the author of *Old Barn Plans* (Main Street Press, 1979), several types of notching were used to form secure corners on the log barn (also called a crib, as was the log cabin). The Swedes and Finns favored tightly fitting double-notching, whereby the logs were dovetailed into one another. In this, they followed a tradition that goes back to the time of the Vikings and the farmsteads of the High Middle Ages, which consisted of a number of log buildings serving various purposes.

At the open-air museum at Skansen, near Stockholm, a farmstead dating from 1574 has been replicated. It has the characteristic two-story, log-built balconied granary, similar to the German forebay. Triangular pieces of timber form a staircase leading to the gallery. Freestanding granaries, built on tall posts to protect the contents from dampness and rodents, were used in many parts of Europe in modified forms. Before the days of pit storage and silos for fodder, European farmsteads sheltered their hay in a barrack, that is, a conical roof supported, like the granary, by sturdy posts. This led to considerable spoilage by wind and weather, and immigrants to North America would devise more effective methods of protecting their livestock's fodder through the winter months.

Log barns on the first frontier were roofed with a variety of materials, depending upon what was at hand. Thatch, bark, lengths of hollowed log laid between the top course of logs and the ridge, and hand-split shingles were all employed. Chinking between the logs was filled in a variety of ways, including the familiar mud, or daub; straw; moss; reeds; and clay, where available. Cement would not be employed until the nineteenth century, and then primarily in the West. Popular trees for log building in the Southeast included chestnut, poplar and pine. Sometimes the timbers were debarked and squared after seasoning; in other cases, they were used as they were. Both vertical and horizontal board sheathing were often nailed to the gable ends for additional weatherproofing.

New settlers often built crude log cribs as dwellings while they focused on clearing their land and sheltering their stock. Many of these small gable-roofed structures later served as corn cribs, or as the nucleus of a larger barn. Traces of such buildings can still be seen in Tennessee, the Carolinas, the Blue Ridge Mountains and the Cumberland Gap, on the borders of Kentucky, Tennessee and Virginia. English and Scots-Irish settlers were quick to adopt the log-cabin model that was so well suited to the milder climate of the lower Mid-Atlantic and Southern regions. Here, gabled roofs often had a more moderate slope than those farther north, because heavy snowloads were not a major problem. Lean-to sheds were added on one or both sides of the log barn to accommodate wagons and tools, and modest porches along the same lines became common where the climate was hottest and most humid. In fact, the ubiquitous front porch is a Southern contribution to North American architecture, influenced by the building traditions of African Americans brought to the region as slaves to farm the labor-intensive plantation crops of rice, tobacco, and later, sugar cane and cotton.

Many log barns were built in the form called "two pens and a passage"—one or two cribs on either side of a wagonway, all covered by a single roof. One of the cribs would be used for grain or hay storage while the other was used as a livestock shelter. Also called the "dogtrot" barn, this simple model was widely used throughout the southern Appalachians and later appeared west of the mountains in the form of the ranch house, as well as livestock shelters built for sheep and cattle on the nineteenth-century frontier.

Very large log barns may still be seen in parts of the Southeast region. They most frequently take the form of a second-story loft raised above two sets of cribs, often cantilevered over the foundation on all four sides and roofed with wide eaves. The loft has a hay door and winch at one end for lifting the hay from wagons at ground level. Where siding appears on the loft, it is usually vertical planking.

What would eventually become the state of Georgia was founded in 1732 by a group of altruistic young English aristocrats headed by James Oglethorpe. They named this last unclaimed tract on the Atlantic Seaboard for King George II and chartered it as a trust, planning to establish a refuge for London's destitute as an alternative to debtor's prison. The indigents were to grow hemp, flax and grapes and to augment their income by the production of raw silk, all of which would be purchased by the trust at fair market value. Slavery was prohibited in the new colony, which the Mother Country envisioned as a potential buffer zone between the Spanish settlements in Florida and the English colonies farther north.

One early colonist wrote home praising Oglethorpe and expressing high hopes for the settlement made at present-day Savannah, despite the inconveniences of runaway hogs, ants that "bite desperately" and a few "Grumbletonians" who found fault with the climate and situation. Unfortunately, their numbers swelled, as the would-be farmers discovered that many of their allotments were either too sandy or too swampy to grow the designated crops for the trust, much less the foodstuffs they needed. Boundary lines were not clearly marked, and straying livestock took a heavy toll on crops. Somehow, plows had been left behind in England, and only a handful of the first 150 colonists had any farming experience. Spanish incursions from Florida proved to be a greater threat than the founders had anticipated, and the colony's agricultural experiments were often thwarted by the need to leave their fields and livestock to fight off the intruders. Within a few short years, Oglethorpe's idealistic hopes of keeping the colony free of slavery were opposed successfully on financial

Previous pages: **Red barn near Knoxville, Tennessee;** *Above:* **Old barn in northern Kentucky**

grounds, and cotton, cultivated by slave labor, became the region's principal crop. The yeoman farmer envisioned by the founders would become a tenant farmer, eking a meager living from the red clay soil and keeping a few cows, chickens and pigs for family use.

During the early 1800s, log barns set on low stone foundations became the norm in Georgia. Cracks between the logs were covered with narrow slats or riven clapboards. Long slats were used like shingles to cover the roof from eaves to peak, nailed down and often chinked with sod or moss. Various outbuildings served the functions of smokehouse, summer kitchen, corn crib and wagon shed, as needed. In this hot climate, as elsewhere in the South, mules were often employed for draft use instead of horses and oxen. They were inexpensive, easy to maintain and more tolerant of the heat than the larger animals. Mules are hybrids, bred from the horse and the ass, and differ in size and strength according to the predominance of the

parental species. Hybrids from a male ass and a mare are superior to those from a she-ass and a horse, which are sometimes called "hinnies" to distinguish them from the other mules. Sure-footed and capable of enduring great fatigue, mules would also play a prominent role on the Western frontier as beasts of burden and draft animals, especially in mountainous country.

No discussion of agriculture on the first frontier would be complete without mention of the several Utopian communities, including the Amish and the Shakers, who contributed so much in proportion to their numbers. The Amish had their roots in the Protestant Reformation, among the Mennonites of Switzerland, who were disciples of a former priest named Menno Simons. He and his followers were also referred to as Anabaptists, because they opposed the Roman Catholic practice of infant baptism and the unity of church and state that had produced so many abuses during and after the Middle Ages. Persecuted

Above: Huntington, Maryland; *Opposite:* The Hermitage Barn, home of Andrew Jackson, Nashville, Tennessee

for their separatism, many Swiss Mennonites moved into the German Rhineland and some migrated eventually to the United States and Canada. An offshoot of this sect was the Amish, who settled in eastern Pennsylvania, especially in Berks and Lancaster Counties. Led by Bishop Jacob Amman, they brought both Swiss and German timber-framing techniques to the New World and built many of the handsome banked barns described and illustrated in chapter 2. German immigrants to the rich farmlands of Pennsylvania were referred to collectively as "Pennsylvania Dutch," derived from the word *Deutsch*, meaning German. Although the Amish communities were very close-knit, and differed from their neighbors in such matters as worshipping in their homes rather than in churches, they were, and are, widely respected for their industrious ways and the prosperous, well-kept condition of their farmsteads, crops and livestock.

Amish family farmsteads grew to include two or three generations in close proximity to one another, and new communities were eventually founded in Ohio, Indiana, Iowa and other areas. The Old Order Amish have adhered closely to their traditional ways of life to the present day, using horses for farm work and transportation instead of tractors and cars. Amish buggies, often drawn by Standardbred horses, bred for light road work and as trotters, are a familiar sight on the roads through their settlements. Belgian and Percheron draft horses may be used in four-hitch teams for disking and plowing. Most families preserve several hundred quarts of meat, fruit and vegetables for winter use and keep enough chickens to meet their own needs and sell surplus eggs. Dairy farming is a major activity, but no electrical appliances, tools, or machinery are used in the stable or the farmhouse. Amish housewives may churn their own butter and bake their own bread; family barns are models of cleanliness and good order. Many have so-called hex signs painted on them, which were long believed to serve as protections against misfortune or witchcraft, but scholars have proved that these colorful designs serve no purpose except ornamentation. They are a reflection of the old Swiss and German traditions of decorating the house end of the *Loshoes*, or house/barn, with bright colors, attractive motifs and carvings.

The Shaker sect originated in England during a revival among the Society of Friends, when foundress Jane Wardley and her husband, James, exhorted their Quaker congregation to purify their way of living. The ideal they proposed included communal property held by a "spiritual family" founded on the type of the natural family, but committed to celibacy. They retained the Quaker principles of nonresistance and pacifism and called themselves "The United Society of Believers in Christ's Second Appearing." Others named them the Shaking Quakers, because of their physical manifestations of spiritual power while engaged in public worship, including liturgical dancing.

An early convert, Ann Lee, was imprisoned for her beliefs and received a vision directing her to emigrate to the New World. She and eight followers settled near Albany, New York, at present-day Watervliet, in 1776. "Mother Ann's" inspired preaching gained many converts, and after her death in 1784, Shaker communes were founded by Joseph Meacham and Lucy Wright, beginning at Mount Lebanon (later New Lebanon), New York. There, before the Civil War, the Shakers built a great stone barn five stories high and 300 feet long, covered by a slate roof. Originally, the gable end was square-topped and suffered water damage, so an alteration was made to the roofline. The central entrance door to this fortresslike building, called the North Family Barn, is at the third-floor level, adjacent to the road. The Shaker craftsmen were entirely self-taught, and no blueprints or drawings have been found that reflect the plan of the building.

As early as 1830, Shaker barns had been admired and described in agricultural journals. In that year, the *Newengland Farmer* reported that "The Shakers of Harvard [Massachusetts] are building a barn which is probably longer than any structure of this kind on this continent. The dimensions, as we are informed, are one hundred and fifty feet in length, and forty feet in width. It is four stories in height, and the calculation is to drive in on the upper floors, from the hillside, and pitch the hay down, thus rendering much hard labor easy." At their community in Enfield, New Hampshire, in 1854, the Shakers built a three-story barn with a high-banked ramp at the gable end and stabling for the dairy herd at ground level, which was illuminated by closely spaced windows.

Perhaps the best-known Shaker building in the United States is the great round stone barn at Hancock, Massachusetts, which was first built in 1824. In his book *An Age of Barns*, historian Eric Sloane makes a persuasive case for the Shaker affinity for the circle as the perfect form—an affinity shared by the Quakers and the so-called Holy Rollers. He observes that "Farmers made circular designs on their barns, and their wives sewed circular patterns on quilts. The Shakers used the circle in their 'inspirational drawings' [comparable to the Eastern mandala]....They took delight in round hats, rugs and boxes; and they made round drawer-pulls and hand-rests for their severely angled furniture." Certainly, this monumental stone barn, carefully preserved in Hancock Shaker Village, approaches the ideal in its unity of form and function. Eight sturdy posts support the large cupola and the radiating trusses that help stabilize the structure. Clerestory windows, now boarded up, admitted light from above, and a circle of posts with parapets forms an aisle around the central mow. Here the cattle were stabled, facing the mow—almost 50 feet in diameter—from which they were fed. Manure basements located below the stalls kept them clean and provided fertilizer for the community's crops.

Completed in 1865 on the foundations originally laid in 1824, the round Shaker barn excited great interest in the agricultural community, and during the latter half of the century, many round, octagonal and polygonal barns were built in both wood and stone. Increasing enthusiasm for "scientific agriculture" helped popularize these unusual forms. Eric Arthur and Dudley Witney cite the publication issued for the

Melrose Plantation barn (1845), Natchez, Mississippi

Friends of Hancock Shaker Village on the opening of the restored barn in 1968, written by Eugene Dodds:

"As progressive farmers on the Great Plains were advised, so they built, and during the last two decades of the century, timber variants of the Round Stone Barn appeared nearly everywhere along the western frontier. A number still survive, especially in Kansas, Nebraska, and the Dakotas, providing to this day the tribute of imitation to what an agricultural writer of the mid-century called 'the superb ingenuity of the Shaker builders of Hancock, whose circular barn should always stand as a model for the soundest dairying practices.'"

Like the Mennonites, the Shakers maintained good relations with the larger community, and their products, from furniture to herbal remedies, were valued. An advertisement from 1890, by A.J. White of 54 Warren Street, New York City, is illustrated with a drawing that is captioned "Shakeresses Labeling and Wrapping the Bottles Containing the Shaker Extract of Roots, or Siegel's Syrup." Mr. White (or his copywriter) announces that the women pictured are members of the Shaker community at Mount Lebanon, New York, and notes in the florid style of the day, with italics for emphasis, that *"The Shakers would not let their good name be used on the medicines if they were not genuine."*

Unfortunately, by this time, membership in the Shaker "spiritual families" had begun to decline, since their rule of celibacy limited their numbers to converts. While their missionary efforts had been successful in Kentucky, Ohio, Indiana and the Eastern states where they first settled, attrition reduced their numbers steadily, and by the late twentieth century, only a few elderly survivors remained to uphold their traditions of purity, simplicity and the expectation of a new millennium conceived and nurtured in peace.

An Architectural Landmark *Below*

The great Shaker barn at Hancock, Massachusetts, was built during the 1860s at the then prohibitive cost of $10,000. The commune's highly skilled masons and carpenters brought all their talents to bear when they created this widely admired round dairy barn on the foundations of an earlier building. The rooftop monitor, with its clerestory windows, and cupola were added during the 1870s after the original roof and interior were completely destroyed by fire. This epochal barn and its outbuildings are preserved in scenic surroundings at Hancock Shaker Village.

Ample and Efficient *Above*

Round and polygonal barns can be found in rural communities across the continent, their antecedents being the European grain mills in which roped oxen plodded around a grinding stone, slowly turning the stone. Multisided, or polygonal, barns like this example in Waitsfield, Vermont, were easier to construct than a perfectly circular barn, which demanded precise engineering. Orson Squire Fowler popularized the octagonal form during the 1840s in his book *A Home for All* (1848), claiming that it provided more floor space, greater economy of movement in performing various chores and more efficient ventilation by way of the rooftop cupola.

An In-barn Corncrib *Below*

Amish settlers who migrated to Ohio from Pennsylvania brought the practice of drying corn for fodder in a loosely constructed crib inside the barn, rather than in a separate V-shaped structure raised above ground level.

Saltbox Roofline on an Amish Farmstead *Above*
This well-built Amish barn in Kidron, Ohio, is used
mainly for hay storage, as indicated by the near-
absence of windows. The haystack has become a
curiosity in most areas, where gasoline-powered
machinery has replaced the traditional hay-wagon
driver, stacker and presser.

Quiet Fields Blurred by Snow *Overleaf*
A Standardbred road horse paces through a winter
scene in Ohio's Amish country, which is centered
around Holmes County.

A Maryland Tobacco Barn *Opposite*
Hinged siding that can be opened or shut as needed
to cure the leaves hung to dry inside are typical of
tobacco barns from the Tidewater states to the
Connecticut Valley.

Southeastern Barn with Brick Silo *Below*
This gable-end barn near Rocky Mount, Tennessee,
has a well-built brick silo of nineteenth century
vintage. The mellow colors of barn and silo blend
harmoniously with the landscape.

Weathered Working Banked Barn *Opposite*
Mount Harmony, Maryland, is the site of this well-
worn, three-story banked barn with a moderately
sloped roof that attests to the region's relatively
mild winters.

Giving Way to the Elements *Below*
An old L-shaped barn in New Jersey uses the
second level, rather than the loft, for hay storage
and has windows only in the livestock quarters at
ground level. Missing boards provide ample
ventilation for the hay.

Enlarged Crib Barn *Overleaf*
The historic Caldwell Place barn in Catalooche, Tennessee, veiled in morning mist, has a large wagon drive flanked by three squared-log cribs. Widely spaced diagonal boards frame the long hayloft above. The roof has lost much of its original wooden shingling to wind and rain.

Stripping Down *Above*
This venerable barn in North Carolina's Catawba
County is being reduced to its sturdy timber
armature, as siding succumbs to disuse and neglect.

Florida Frontier Barn *Overleaf*
The widely spaced boards of this ramshackle barn in Columbia County, now overhung by Spanish moss on fast-growing trees, attest to the peninsula's subtropical climate. Florida was very sparsely settled even in the late nineteenth century.

Kentucky Tobacco Barn *Below*
The open-air tobacco barn in the central Bluegrass region combines form and function effectively. Dressed-log uprights with cross-bracing boards provide ample ventilation.

Expanding with the Times *Above*

This sprawling complex in Lambertville, New Jersey, shows that agriculture remains a major industry in the Garden State, first settled by skillful hard-working Dutch farmers in the seventeenth century.

A New Kind of House/Barn *Overleaf*
Historic preservationist Marty Azola, his wife,
Lone, and hard-working friends and colleagues
joined forces to turn a dilapidated dairy barn near
Baltimore into a harmonious home. It incorporates
a horse stall-turned-study, a cow-washing room
transformed into a master bedroom and a breezeway
enclosed to form a sunny breakfast nook.

A Flourishing Maryland Farm *Below*
The years have been kind to this trim red-and-white
dairy farm in Easton, Maryland, with its gable- and
gambrel-roofed barns, fenced farmyards and
spacious sheds and silo.

Peaked Roofline with Extensions *Right*
This large early-twentieth-century barn in Lewes, Delaware, has a lean-to extension on the right side and a projecting shed with a small peaked roof on the left. Entry is from the gable end.

White Angles and Planes *Below*
This modernized farmstead in south-central New Jersey, with its several neat outbuildings, is a serene presence in the early-fall landscape.

Tobacco Curing, Tennessee *Overleaf*
Unpretentious, but still serviceable, an old farm barn
in Tennessee's Dry Hill Community is used as it was
a hundred years ago.

Barns of
Eastern Canada

Previous pages: **French River, Prince Edward Island;** *Above:* **Shingled outbuildings on a typical Maritime Provinces farmstead**

The earliest history of French barns and houses in North America has been preserved in the archives of the province of Quebec, beginning in the early seventeenth century when the first permanent settlement was made by Samuel de Champlain in 1608. Those who came from France to settle in the fortified village of Quebec were called *habitants*, to distinguish them from the many French fur traders, soldiers and government officials who visited and then returned to Europe. The rich alluvial soil of the St. Lawrence River Valley, once cleared of timber, yielded abundantly, and among its first farmers was Louis Hebert, who grew grain and vegetables near Quebec from 1615.

The region's frigid winters demanded sturdy shelter for both people and animals, and the first farmsteads in New France drew upon the building traditions of Normandy, Brittany and other regions of the Old Country, where timber had been growing scarcer since the late Middle Ages. In the New World, elm, ash, oak and other trees grew abundantly, and both timber and fieldstone were used for construction. Sill beams were usually laid on a masonry foundation, and closely set posts, or studs, were infilled with wattle and daub, fieldstone, cobs and other weatherproofing materials. The familiar European house/barn plan had been modified in France to form the *maison rudimentaire*, which in Normandy became a form of connected architecture encircling a courtyard—the *maison cour*. This plan was widely adopted in New France because of the degree of shelter it afforded and the convenience of having various storage and work areas adjacent to one another. As modest colonial farms grew larger, the farmyard might comprise a woodshed, poultry house, dairy, tool shed, sheepfold and other structures under separate roofs.

French settlement extended south of the St. Lawrence into the arable land now known as the Eastern Townships and north of it to the good farmland in the neighborhood of Three Rivers, some 90 miles from Quebec. Originally, the farmland along the St. Lawrence, the gateway to New France, was allocated to various *seigneurs* by grants from the king. The parcels were divided into long, narrow holdings that fronted on the river—almost 200 feet long and perhaps ten times deeper, taking in pasture and woodland farther from the waterway. Tenants held these lands along feudal lines, paying rent in perpetuity, with certain goods provided to the *seigneur* on an annual basis, usually in the form of livestock or grain. A pair of geese or chickens, or a bushel of wheat, might be contributed by the tenant, who also agreed to grind his grain at the landholder's mill, where one-fourteenth of it was taken in compensation as *moliere*, or toll for grinding. In fact, these conditions prevailed in Quebec long after its conquest by the British in 1763. Not until 1854 were the old laws commuted. Until that time, the rental of riverfront land might be reckoned at "twenty sous and a good live capon" per area of frontage, while a 130-acre farm was leased for 80 sous and four capons.

Meanwhile, French settlers had founded what is now the island city of Montreal (formerly Ville Marie) and Acadia, along the Atlantic Coast, centered on present-day Nova Scotia. Missionaries, notably the French Jesuits, to the Huron and other native peoples pushed west into Ontario, where they founded the mission of Ste. Marie to the Huron in 1639. This frontier outpost of the "Black Robes" was of rugged log construction, surrounded by a palisade. It had a two-story barn with narrow windows flanking the doors to the stable area, which is protected by wide eaves under a sloping roof designed to shed the region's heavy snowloads. Fortunately, this historic barn has been rebuilt for posterity using the methods employed by the missionaries Jean de Brebeuf and Gabriel Lalemont during the seventeenth century. Here they harvested Ontario's first cereal crops, including barley and oats, and experimented with the cultivation of native corn and sunflowers, all of which would figure in the future of Canadian agriculture.

Long called Lower Canada, the Province of Quebec has been described as having "the summer of Paris and the winter of St. Petersburg." Annual snowfalls up to 150 inches have been recorded here. Thus the first *habitants*—a word that became synonymous with country people—built sturdy rectangular barns of wood and earth walled with *gasparde*, or wattle and daub. Originally bonded with mud, and later with plaster, these time-honored materials protected oxen, cows and horses from the rigors of winter. Most families also kept some poultry: chickens for eggs and meat, and geese for their eggs, feathers used for bedding, and the traditional roast goose and paté. Pigs, sheep and beef cattle were also imported and would eventually be bred selectively to improve the quality of Canadian livestock. By 1782 the author Hector de Crevecoeur could observe with a note of pride: "For my part, I had rather admire the ample barn of one of our opulent farmers, who himself felled the first tree of his plantation and was the first founder of his settlement, than study the dimensions of the temple of Ceres."

Perhaps less ample, but no less essential to the agricultural history of Canada, were the long farm buildings modeled on the house/barn of the traditional Breton farmstead. Unlike the Norman *maison cour*, the *maison bloc* housed all facilities under a single roof. The house end of the building had its own entrance and larger windows than the section that sheltered livestock and implements. A dormer door near eave level received hay and grain into the upper storage area, which was filled by hand from a ladder or stone stairwell. As in Spain and France, a cross, or an image of Christ or the Virgin Mary, was often placed in the barn for protection and blessing of the granary and the animals.

According to Canadian historians Eric Arthur and Dudley Witney, authors of *The Barn: A Vanishing Landmark in North America* (Arrowood Press, 1989): "Even without the *maison* [house], the Breton form remained in the great nineteenth-century barns. The dormer evolved as a dominating and striking element in the façade, and served as a wagon entrance to the left of the stable." By their account, based on original documents, forty-seven house/barns of the authentic Breton type were built in the region of Quebec,

Montreal and Three Rivers between 1662 and 1771. These connected barns included one formed by the merchant Pierre Renthuys of Montreal in 1694, when he rented "half a stable which he added to the end of his house in Notre Dame Street to winter his animals."

Except for the log barn, in which hewn logs were slotted into upright studs in the manner called *pièce sur pièce*, or one on another, most Quebec barns were whitewashed, including the later buildings faced with vertical planking. Their long, low silhouette told its story of barn, stable, cattle byre and wagon shed within. The stone barn that had served well in France proved unsuitable in the severe cold of New France, where heavy frost formed on the inner walls to the detriment of the animals' health and comfort. Over time, thatched roofs were largely replaced by overlapping planks or cedar shingles. Four-sided hipped roofs were often built, but steep gable roofs—some with projecting peaks or flared eaves—were more common. The bell cast of the flared eave was also seen in New Netherland long before the American Revolution; it is believed to be of Flemish or Walloon origin. Later, recessed walls and doors on the German model were sometimes added to Quebec barns for additional weather protection. Here, a large archway gave access to the drive floor and the inner stableyard.

Eighteenth-century Ontario barns were often built with chinked logs and a shingled gable roof. Small entrances at either end flanked the main entrance. The three-sided, or U-shaped, barnyard was common, centered on the wheat barn with its large threshing floor, one or more granaries and a sizable mow for the sheaves. Above the threshing floor was the loft, supported by a timber called the swing beam—a New World innovation. According to Arthur and Witney, "By the end of the eighteenth century in England, few barns were built of wood, and the majority would be of brick or stone. Mr. Anthony Gervase [author of *Architecture and Town Planning in Colonial Connecticut*] is of the opinion that the technique of frame construction as we know it, in the barns and domestic architecture of North America, can be traced to the earliest settlers of New England. Essex [a county in southeast

Bow-truss-roofed red barn, Annapolis Valley, Nova Scotia

England with a long tradition of building in wood] was the birthplace of many of these immigrants."

Along one side of the barnyard was a cattle shelter with hayloft above, a cow shed (byre) with half a dozen or more stalls and a sheep shed with a manger along one wall. Opposite these buildings was a large stable barn with a ground-level hay mow in one bay and stalls for the horses, with feed bins and tack wall, in the other. The central bay was a drive floor that doubled as a storage area for wagons and other implements.

Scottish immigrants to Ontario, well known for their masonry skills, built stone barns and farmsteads. One example, now fallen into ruin, is a farm near the town of Arkell. Its courtyard had stables on the left and a byre on the right of the main barn, whose arched entry bears the carved datestone "1871."

Before and during the American Revolution, many Loyalists left the New England colonies to settle in what had recently become British Canada. Some moved into Ontario, where they built the three-bay English barns described in chapter 1 and called generically "New England style" in Canada as well. Ontario's Upper Canada Village re-creates a British community of the 1800s on a 66-acre site that includes two farms, three mills and several churches.

Other colonial Loyalists crossed into Atlantic Canada, comprising the present-day provinces of New Brunswick, Nova Scotia, Prince Edward Island and Newfoundland, including Labrador. Some 3,000 of them arrived in Nova Scotia aboard thirty ships from New York City and founded the town of Shelburne, with the active encouragement of Canadian authorities. Quaker pacifists also left the embattled American colonies to settle here as farmers, whalers and merchants. Barns and outbuildings from the Loyalist era have been preserved at Kings Landing Historic Settlement, west of Fredericton, New Brunswick.

Much of this region, originally called Acadia, had been settled by French immigrants, who fished and farmed here from the early 1600s. After the area was claimed by the British in 1713, during Queen Anne's War, the Acadians resisted British rule for decades. As early as 1785, French-speaking Acadian farmers from what is now Nova Scotia were forcibly relocated. Many

emigrated to present-day Aroostook County in northern Maine. The barns they built there combined French and English influences, with shed-roofed stables along the rear of the building and hip-roofed sheds at the gable ends. Other Acadians moved west and followed the Mississippi, where the French had constructed a chain of forts to protect their claim to the region, migrating as far south as the bayou country around New Orleans, where they were known as Cajuns. Eventually, some of the Acadians returned to their homeland, where their descendants live in areas including the Evangeline region of Prince Edward Island. Their saga was eloquently memorialized by Henry Wadsworth Longfellow in his epic poem "Evangeline."

Another casualty of British conquest was the historic French fortress of Louisbourg on Nova Scotia's Cape Breton Island, which dominated its Atlantic site from 1713 until 1758. Fortunately, Canada's largest historic reconstruction project has been carried out here to replicate much of the old town and its adjacent fortress. The seaside location still poses challenges to agriculture, but vegetables that were grown here in raised beds 200 years ago are being planted again today, including cabbages, turnips, scarlet runner beans, Blue Pod beans, herbs and Spanish radishes. Manure from the barnyard and wood ash for pest control make Fortress Louisbourg an example of organic agriculture as practiced several centuries ago. Chicken breeds that have now become rare outside of agricultural exhibitions, including Hamburg, Black Polish and Cochin, lend colorful notes of black, white and red as they range freely outside their coops. Hearty French loaves are baked as they were for the soldiers' weekly ration when this settlement was one of the three major ports on the Eastern Seaboard.

The nineteenth century brought another wave of immigration from the British Isles to Eastern Canada, along with emigrants from Europe and the United States. Many of these newcomers farmed side by side with the long-settled inhabitants and introduced new forms of vernacular architecture. The Hutterites, for example, were a German sect of Anabaptists who emigrated from Moravia and the Tyrol to the New World. They were named for their leader Jakob Hutter, who

Above: **Barn and windmill, Ontario;** *Opposite:* **Land mist in Tea Hill, Prince Edward Island**

had been martyred as a heretic in 1536, and founded collective farms called *Bruderhöfe* in colonies of up to 150 people. Like the Amish and other Mennonite sects described in the previous chapter, the Hutterites maintained their own way of life, building log structures and house/barns based on central European and Swiss designs. Both Hutterites and Mennonites still farm regions of Ontario, including those centered on the villages of St. Jacobs and Elmira.

The country people of Quebec also adhered to many of their traditional lifeways, raising mainly hay, oats and potatoes. The 1897 book *Canadian Folk-Life and Folk-Lore* by William P. Greenough, republished in a facsimile edition by Coles in 1971, offers a fascinating glimpse of rural life in the province at the turn of the century. Livestock included dairy cows, pigs, sheep and poultry, while beef cattle were purchased from the

grain-growing province of Ontario, or from the Eastern Townships bordering Maine, New Hampshire and Vermont. Traditional foods remained popular among the *habitants*, including six-pound loaves of French bread, boiled pork, sausage, game birds, seafood, pea soup, apples, strawberries and maple syrup, for which sap was boiled in a "sugarhouse," as in New England. In France and the Low Countries, large working dogs had been used as draft animals to draw carts filled with milk, wood and other provisions. In the New World, they were used to the same purpose as sled dogs.

Sheep like the French Rambouillet, with large, rugged conformation and heavy fleeces, set the standard for both wool and mutton. Originally called the French Merino, such sheep had been established as a breed in the Royal Flock of Louis XVI, who imported almost 400 of the best Spanish Merinos to

France to foster the wool industry. Major cattle breeds of French origin include the white Charolais, bred for strength and utility rather than refinement. Legend has it that wild white cattle were first domesticated during the Middle Ages in the old provinces of Charolles and Nièvre. Heavy-bodied and muscular, they were prized for meat, milk and draft use and selected for traits including rapid growth and large size. The breed became important in the United States during the nineteenth century, and importations were made from France by way of Canada. Later, Western cattlemen would crossbreed the Charolais with the hump-backed Brahman to produce hardy beef cattle resistant to hot, humid conditions.

Canadian farm horses were bred for economy, strength and endurance, like the French Percheron, the Belgian and the English Shire horse. They were generally short-legged and heavy-bodied, with broad chests and large heads, able to work hard on indifferent feed. Sometimes they were used in tandem with oxen for field work, and they hauled travelers and produce over bad roads in every weather. In 1897 Mr. Greenough was deploring the fact that "the race of Canadian horses that was famous fifty or seventy-five years ago has entirely disappeared, and its equal for speed and hardiness has not been found." Shortly after this time, North American farmers would start to measure "horse power" against newly developed machinery and internal combustion engines that marked a new, mechanized era in agriculture.

The bonds between Eastern Canada and New England remained strong, and as many American farmers moved west in search of new opportunities, some Canadian farmers migrated south. In a statement that many a hard-working Yankee farmer would have contested, Greenough observed that: "To some men of the younger generations of these *habitants*, abandoned farms in New England have seemed to offer greater temptations than their native country could show them. The number of these farmers is not very great, but I understand that such as have taken such farms have almost invariably been successful. Patient and frugal, they are content with results that did not satisfy the more restless and ambitious Americans."

Pioneer Chinked Log Barn *Above*

During the eighteenth century, many Ontario barns were built of chinked logs, like this rustic example from Pelee Island, at the western end of Lake Erie. Almost the southernmost point of Canada, at the same latitude as northern California, this part of Ontario has a lower annual snowfall than much of Eastern Canada, as evidenced by the roomy loft space below a moderately sloping roofline: Elsewhere, barn roofs are usually more steeply pitched.

A Fieldstone Classic *Opposite, above*

Many early Scottish immigrants to Ontario built farmhouses and barns of stone. This beautifully constructed barn is at Upper Canada Village, where a late-eighteenth-century community has been re-created with historic buildings relocated from nearby villages that were submerged when the St. Lawrence Seaway was altered during the mid-twentieth century.

Old-time Sheep Farming *Opposite, below*

Thunder Bay, Ontario, the location of this historic farm/barn building, is Canada's second-largest port and the world's largest grain-handling center. The area is home to the descendants of immigrants from around the world, including the largest Finnish population outside Finland. Now converted into a farmhouse-style dwelling, this wooden structure shows northern European influences in its design.

Good Earth: Prince Edward Island *Opposite*
This picturesque barn in Brookfield, with its undulating furrows and fence line, reveals the character of the island's rolling plains. Until recently, it had the highest proportion of farmland of any Canadian province. Claimed for France by Samuel de Champlain in 1603, Prince Edward Island is typical of Eastern Canada in its diverse barn styles— French, English, Scots-Irish and others—as seen in the following plates.

The Connected Barn *Overleaf*
Continuous architecture on this farmstead indicates several generations of occupation and progressive enlargement, as the rich red soil produced the crops for which Atlantic Canada is renowned.

Sturdy Well-fed Cattle *Below*
The white-faced Hereford, a breed that originated in England, became a mainstay of the Canadian livestock market.

The Celtic Influence *Above*

This barn and shed in Irishtown, Prince Edward Island, recall the whitewashed barns and cottages of Ireland, with their sloping roofs (originally thatched) and small windows. The hay storage area in the upper level is the largest section of the barn, which has a lean-to addition in the saltbox style that was widely used in New England.

The English Barn Writ Large *Opposite*

This venerable barn was once the pride of a British Canadian farmer. Its timber framework has withstood many a severe Atlantic winter, but the ridgepole is sagging now, and the hayloft open to the elements. The wind blows through the central bay on the eaves side and the wagon drive at the foot of the hill.

Pennsylvania-style Banked Barn *Overleaf*

The German and Swiss antecedents of this banked barn are apparent in its siting on the hillside and the grass-grown earthen ramp leading into the central bay. Only the ground-level livestock area has windows. The field of lupines in the foreground sounds a colorful note in this pastoral scene.

Idyllic Pasturelands *Above*
This well-kept dairy farm, with its view of a green
patchwork quilt of farmland and woods, has twin
barns to house its herd of Holstein-Friesians and the
operations of the dairy. Brightly painted trimwork is
a feature of many Eastern Canadian barns, especially
in regions of Scots settlement.

A Study in Contrasts *Opposite*
Gambrel-roofed, black-and-white barns, with a
neatly fenced farmyard and outbuildings, stand
serenely under an immense sky flecked with clouds
at St. Ann's, Prince Edward Island, originally settled
by the French.

Winter's Rest *Overleaf*
A modest English-style farmstead in North
Wiltshire, Prince Edward Island, is the only sign
of human presence in the pristine snow-covered
land- and seascape of the island that the native
Micmac called *Abegweit*: "moored in the shelter of
the encircling shore."

Whitewashed Barn and Outbuildings *Above*
White barns with contrasting trimwork—this one an
English-style barn with lean-to addition—have been
a familiar sight in Eastern Canada since the time of
French settlement. Even log barns were sometimes
plastered over for additional insulation; frame barns
were whitewashed with a mixture of lime and water
and, where available, powdered chalk called whiting.

A Snug House/Barn *Opposite*
This isolated farmstead under a blanket of snow
shows the advantages of connected architecture in
regions where winters are severe. The steep roofline
on the almost windowless barn efficiently sheds
heavy snowloads, under which a moderately pitched
roof might buckle.

An Abandoned Farmstead *Above*

This once-flourishing farm in China Point has fallen victim to the growing trend toward consolidating Prince Edward Island's many small farms into larger holdings. Until recently, the island was almost entirely rural: Residents refer to it proudly as "the Garden of the Gulf."

Signs of Late Summer *Opposite*

Hay bales ready for winter storage dot the fields of this modest English-style farmstead. Livestock fodder and potatoes are among Eastern Canada's principal crops.

The Saltbox Style *Overleaf*

A saltbox barn, sloping "into the weather" on the north side, is a rural building style common to Eastern Canada and New England. This example is paired with a spacious gambrel-roofed barn with a large storage capacity in the loft level and livestock stalls below.

Smokehouses on Stilts *Left*
Commercial fishing is a major enterprise in the province of New Brunswick, both for local consumption and export. These smokehouse barns at Seal Cove, on Grand Manan Island in the Bay of Fundy—the southernmost island of New Brunswick—are kept safe from the tides by sturdy pilings driven into the shoreline.

Farmers and Fishermen *Overleaf*
The flared eaves on these storage barns lining Malpeque Bay, on the northern coast of Prince Edward Island, attest to the region's settlement by the French in the early 1700s. They were followed by many Scots later in the century. Malpeque is famous for its colorful flower gardens, which include hundreds of varieties of roses and dahlias. Before the Europeans arrived, this was a Micmac encampment on the Gulf of St. Lawrence.

Heartland and Great Plains Barns

Previous Pages: Barn (1892), Maplewood, Wisconsin; *Above:* L-shaped limestone stable on the Hutchins farm in Iowa

The poets tell us that Americans are a westering people, perhaps drawn onward by the ever-receding sun. Stephen Vincent Benet reflected: "It's an old Spanish custom gone astray/a sort of English fever, I believe/or just a mere desire to take French leave." Whatever the cause, the early 1800s saw a swelling wave of migration across the Appalachians into the American heartland and the trans-Mississippi West. The first frontier began to feel overcrowded to many homesteaders, who made their way into the Ohio River Valley and lands opened up by the Louisiana Purchase of 1803—some 828,000 square miles—to settle what is now the Midwest. Soon they were joined by growing numbers of European immigrants impelled by poverty, land hunger and widespread political unrest at home. The migrants then spread onto the Great Plains, long considered a kind of desert in the mid-continent. But these vast, almost treeless, prairies were endowed with rich soil for cultivation, once the "sodbusters" had cut through the deeply rooted grasses and cleared away the stones.

What kind of barns and farmsteads did our ancestors build as they settled and worked this fertile land? The answers are almost as various as the countries or regions they came from, and are also shaped by factors like the nature of the land and the patterns of local settlement—the crops that flourished, the livestock raised, access to markets—all of which changed dramatically over the course of the century. However, there are many common denominators across this region, as well. Whether built of logs, boards, brick, stone, sod, or a combination of these materials, these historic barns meet the criteria outlined by Charles Klamkin in *Barns: Their History, Preservation and Restoration:* "The principal aesthetic appeal of an old barn lies in the feeling of essential rightness. Its site on the land, its orientation to the weather, the structural proportions, and the use of materials all contribute to the sense of harmony. We know instinctively that all the elements in its construction blended to create a structure truly appropriate to its surroundings and the purpose for which it was built."

Settlers from Germany, Scandinavia and the British Isles brought their traditional building styles to the

Ohio River Valley and the upper Midwest, with its many lakes and woodlands. Dairy farming became a major activity, as did the cultivation of a variety of grain crops and corn, which was widely used as livestock fodder. Grain bins were actually separate rooms within the barn, tightly constructed for storage of threshed wheat, oats, barley and other cereal crops. These bins had a strong, dry floor, plastered walls and sturdy doors that could be secured against theft and invasion by rodents and other pests. Corn cribs were sometimes incorporated into German and Dutch-style barns, but most were free-standing structures designed to dry the ears of corn for livestock consumption over the winter months. They combined the functions of ventilation and rodent control, so they were usually built with slatted walls slanting outward from base to eaves. This formed a V-shaped receptacle that was raised on posts covered with inverted tin pans to discourage the entry of rats and mice, on the same principle as the mushroom-shaped staddle stones that supported English granaries. The average corn crib measured 6 to 8 feet wide and 10 to 20 feet long.

Another method was to store the corn in a rectangular above-ground bin covered by a shed roof supported on posts. The silage was compressed by rocks in the lower part of the structure. Some corn cribs evolved into full-fledged storage barns, with gaps between the horizontal boards up to the gable level and a rooftop cupola for additional ventilation. One such example is a handsome red corn-crib barn on a fieldstone foundation that has been preserved at Pleasant Hill Shaker Village, Kentucky.

Farming families of French descent occupied the Mississippi Valley from northern Minnesota to New Orleans, where the great river flows into the Gulf of Mexico. Their log barns, as described in chapter 4, comprised upright studs set between a timber sill and an overhead plate and walled with hewn logs that fitted into the studs. In low-lying, swampy areas like Mississippi and Louisiana, these barns were often raised over a stone foundation to prevent the timber sills from rotting. As new settlers from the East moved into the Mississippi Valley, they borrowed freely from the earlier inhabitants and modified their

own Pennsylvania-style, Dutch, log-crib and English barns to meet new climatic conditions.

During this period, the gambrel roof became a standard style, closely identified with North American agriculture across the continent. Its superior storage capacity made it the first choice of prosperous farmers from many different backgrounds. The highly efficient sliding door on rollers also came into general use, along with new types of machinery—driven by the old type of horse power—that lightened the load of farm work, as described in chapter 7. The classic "barn red" remained the most common color for barns that were painted rather than allowed to weather naturally. Before commercial paints became widely available, the economical farmer relied on his tried-and-true combination of linseed oil made from flax, casein from cow's milk and red ochre or oxide from the soil.

As early as 1786, the Ohio Company had been formed in Boston to purchase land and settle New Englanders wanting homesteads in the 1.5 million acres along the upper Ohio River. A year later, the Congress of the newly formed United States passed the Northwest Ordinance to establish a government north of the Ohio. The original "Northwest Territory" included the present-day state of Ohio—a land of good soil and rolling pastureland that attracted many nineteenth-century homesteaders. They came overland and by riverboat, canal barge and the early railroad lines that originated on the East Coast. The Land Ordinance of 1785 had provided that the Northwestern territories be surveyed and divided into six-mile-square townships, each divided into 36 lots of 640 acres. This was a major improvement over haphazard colonial systems, which had allowed for endless bitter disputes about boundaries and property lines.

Many German settlers from the mid-Atlantic region moved into what is now Ohio, including a number of Amish families, who made Holmes County the center of the world's largest Amish community. The Yoder-Miller Farm, founded by brothers Christian and Jacob Schlabach in 1826, now houses three generations of a family that has been here for eight generations. The barn, dating from about 1840, is a typical Pennsylvania-style banked barn, reflecting the family's original

The restored Utopian commune of New Harmony, Indiana (1815)

settlement in Somerset County. It has been added to several times and now includes a straw shed, hog house and silo. Straw is still blown into the straw shed using a "wind stacker" on the threshing machine. The farm's fifteen dairy cows are milked by hand, and the old-fashioned milk cans are picked up every morning by a truck from the cheese house. (Originally, each farmer hauled his milk to the cheese house by wagon.)

The family also sold eggs from their flock of 400 chickens in the nearby village of Charm. The poultry business was begun by Rudy Yoder, who was a courageous conscientious objector during World War I. The government had no provision for conscientious objectors: In fact, he was threatened with execution for refusing to don a uniform. Many members of the so-called peace churches—Amish, Mennonite, Brethren and Quaker—were forced into Army camps. Some of the Mennonites emigrated to Canada, while Amish draft resisters were transferred to Camp Sherman, Ohio, where they worked as cooks and groundskeepers, under a steady barrage of ridicule and contempt, until the war ended and they were mustered out.

As Midwestern farms flourished, they often extended their barns by additional bays and con-structed outbuildings for various functions: spring houses, woodsheds, wash houses, wagon sheds, windmills and toolsheds. The first upright silo has been attributed to an Illinois farmer named Fred Hatch, who built it during the 1870s. It was widely adopted as a superior storage facility for silage—usually corn, which was plentiful and nutritious. The whole plant was chopped up by a feed cutter and blown into the top of the silo through a pipe. Sometimes it was trodden down to expel air and prevent spoilage. Originally, silos were built mainly of wood, as cylinders of thin boards bound by wire hoops. Brick and stone were also used, in both square and circular forms. The convenience of this mode of fodder storage over the old trench method, or the open-air European hay barrack, recommended itself to many farmers, including the New England dairymen, who found that corn increased the milk yield of their cows. By the early 1900s, some half a million silos were in use across the continent.

The Utopian community of New Harmony, Indiana, was founded along the Wabash River in 1815 by the German Harmony Society, led by George Rapp. Ten years later, the Harmonists sold the agricultural commune to Robert Owen, the British social reformer who had made major improvements to working conditions in the cotton mills of New Lanark, Scotland, from 1800 onward. Owens's ideals were based on humanitarian rather than religious principles, and his social order called for community ownership and equality of work and profit.

New Harmony was composed of one- and two-story log buildings based on German vernacular models. Owen made the commune a well-known scientific center, focused on botany and geology. His partner William Maclure, a wealthy American geologist, sent the first seed for the Chinese golden-rain tree to Indiana during the 1820s, and eventually these beautiful trees were planted along the streets of the town. Although New Harmony remained a cultural center until the Civil War, its communitarian principles proved unworkable, with the Owenites splitting into various factions. Some of the buildings erected by the Harmonists and Owenites have been restored by a consortium of Indiana state conservation organizations and the National Society of Colonial Dames. The original Harmonists returned to Butler County, Pennsylvania, where they founded the village of Economy (later Ambridge). Like the Shakers, they practiced celibacy, and the agricultural society died out in the late 1800s.

Notable stone barns were built in the Midwest from the early nineteenth century, including the plastered limestone barn built at Squire Michael Porter's farm in Sharon Township, Michigan. Its massive arched entryway with a decorative surround is at the gable end, and there are an unusual number of windows on both front and side elevations. A moderately sloped gable roof covers a wide central drive floor flanked by several horse stalls and feed bins.

Many Missouri barns had lean-tos on both sides, for stabling cattle and horses. The mid-century William B. Collier barn, in Audrain County, Missouri, appeared as a rendering and floor plan in the 1881 book *Barn Plans and Outbuildings*. An impressive eighty-four feet square, it had generously proportioned stalls, space for wagons and carriages on each side of the central entryway, and storage for hay and grain on the upper level. Three large cupolas for ventilation lined the peak of the roof. The same publication also provided plans for a small two-story barn suited to "the many small farmers, villagers,

gardeners, etc., who wish only barn room enough for a single horse and carriage and a cow." The central ventilator doubled as a chute from which hay and straw could be sent to the ground level, and grain for feeding was housed in bins below the stairway to the loft. This board-and-batten structure had small windows in the gable ends and a hinged hay door on the upper level.

Most woodsheds were attached to another building—house, barn, or smokehouse—in the form of a lean-to, but on large farms and estates, a separate woodhouse was often constructed, open at one end away from prevailing winds. Such outbuildings might also be located on the edge of the woodlot, where firewood was stored in quantity until needed at the main house. An unusually elaborate example is the woodshed at the Adams National Historic Site in Quincy, Massachusetts, divided into three storage bays, each with its own entrance. Stylistic details include ornamental arches and doorheads in the Federal style that was popular during the early nineteenth century.

The mid-century interest in round and octagonal barns extended all the way to the Canadian plains, where many examples remain. Most of these barns were of wood framing rather than stone, and many had an interior silo that projected through the roofline. The largest round barn in Western Canada is at Arcola, Saskatchewan, and a well-worn, two-story structure near Empress, Alberta, is a local landmark known as "Ma Bell" after a previous owner. Another handsome round barn, painted white, with a shingled black roof, is crowned by a large cupola overlooking Rouleau, Saskatchewan. Bob Hainstock, the author of *Barns of Western Canada*, observes that "The round barn stands apart from others, telling us the original owner was a progressive fellow in his time, usually with close ties to American agriculture." Such ties were characteristic of the Great Plains regions, where Americans and Canadians shared similar hardships and opportunities.

Homesteading the Great Plains called for new approaches to shelter for both animals and people. Wood was in short supply, and water was usually provided by a windmill or a hand pump. High winds swept over the flat grasslands, bringing blizzards in the long winter months and frequent tornadoes that demanded the construction of underground shelters. Western pioneers improvised livestock shelters from simple pole frameworks covered with hay or straw while they constructed more permanent barns. Dwellings and stables were also dug into hillsides and roofed with sod, which provided good insulation from winter's severe cold. However, such roofs leaked mud into the shelters during heavy rains and were sometimes blown away in high winds. Constant vigilance and back-breaking labor were the norm for the pioneer prairie farmer. In Saskatchewan, Ukrainian immigrants who had no draft animals yoked themselves to the plows to break up the hard-packed topsoil for planting.

Several new barn styles evolved as the Great Plains were brought under cultivation. One had a tri-pitched gambrel roof, with shallower pitch over the wings extending from either side. The ground-floor level was devoted mainly to livestock housing and feeding, with storage for wagons and machinery on one side and a hay door with winch at the roof peak. Another type of prairie and Far Western barn had a two-story hay storage center with open peaks for ventilation. The openings were protected by projecting rain hoods, which

Polygonal barn with windmill, Union City, Iowa

led to the nickname "top hat" barn. Single-story lean-tos on each side housed the livestock, and hay was loaded mechanically into doors high in the gable ends. The prairie corn barn was a single-story building with a wagon door at one or both ends. Once threshing by hand had been replaced by mechanical methods, the former threshing floor was called the drive floor and served to house wagons, tools and machinery.

By the end of the century, timber framing had been almost entirely replaced by new methods of barn building developed by improvements in both technology and transportation. One was called stick-frame construction and used wood cut to standard dimensions by a sawmill. Light walls were supported on upright studs, and the roof was carried on board frameworks called trusses, which replaced the old post-and-beam sections pegged together by hand. From the burgeoning city of Chicago, the new mid-continent railway and building center, came the technique of balloon framing, whereby small members were nailed together to form the framework rapidly and economically. The studs extended the full height of the building—usually two or three stories—unlike the platform-framing method, in which each floor was framed separately.

During the late nineteenth century, when transcontinental steel tracks had cut a path through the buffalo grass and grain fields of the North American plains, towering grain elevators sprang up along the railway lines. They were built mainly of stacked and nailed two-by-fours to store the huge grain harvests until they could be discharged into freight cars for transport to population centers. Most had monitor-style roofs, resembling a small building set on the roof of a larger one—a style still seen today in old barns and warehouses that have survived the last hundred years. They served the same purpose as a cupola in providing light and air to the topmost level of the building. Sometimes weathered grey, sometimes brightly colored and occasionally roofed or sided in metal, these elevators were majestic landmarks from Alberta and Manitoba to Nebraska, Montana and the Dakotas. A few examples are still in use today, but most have been demolished, and some converted to other uses.

As the Victorian era drew to a close, gentleman farmers, wealthy landowners and architectural writers like Calvert Vaux and Andrew Jackson Downing began to make their influence felt across the country, not only in residential designs, but in plans for stylish stables, barns and outbuildings. Downing's *Cottage Residences* (1873) included renderings and floor plans for stables, gazebos and estate barns in the "Rustic Pointed [Gothic] Style," and venerable three-bay barns sprouted bargeboards and window hoods in the lacy Queen Anne mode. Some architect-designed estate barns could easily be confused with the main house, with their mansard rooflines, double oak doors and Flemish gables crowned by finials. Jaded city dwellers looked to the suburbs and the country for a respite (often highly romanticized) from the pressures of business, noise and traffic. Publishers like Doubleday Page and Company, founded in 1890, capitalized on the ever-growing popularity of country living with dozens of books on gardening, specialized farming, rural architecture and crafts. After 1900 they added successful magazines including *Country Life in America*. The experiments resulting from this nostalgia for the rural life ranged from comical to grandiose, as explained by Clive Aslet in *The American Country House*:

"Some people were tempted by beekeeping, but by far the most popular investment for the man with little land or capital was poultry. What breed to raise? In *How to Make a Country Place*, Joseph Dillaway Sawyer described lyrically how he had tried 'wild squawking brown, also white, Leghorns,…the phlegmatic, good-natured partridge; buff and white Cochins, feathered to their toe-nails; the barred and white Plymouth Rock; the strutting, tufted Poland; the silver penciled Wyandotte; the aristocratic white, buff and black Orpington; the jet black Minorca; the sprightly, trim Rhode Island Reds…and the tiny demure Bantams, who proved more intelligent than their pompous neighbors, notwithstanding the statement that a chicken's education ends when a day old.'" Clearly, this would-be poultry breeder enjoyed the experience, but he was the first to admit that the chicken business cost him much more than it paid. Entrepreneur O.C. Barber had a similar setback when he turned to vegetable growing on his estate at Barberton, Ohio.

Bow-truss-roofed barn, Austin, Manitoba

In a classic case of West meeting East, Marshall Field III, the heir to the Chicago department-store fortune founded by his grandfather, built a huge estate called Caumsett, on the English model, on New York's Long Island Sound. An avid sportsman, Field retained Alfred Hopkins—the leading architect of the fashionable "barn group"—to design a handsome early Georgian-style complex that included stables for hunters and polo ponies; barns for cattle, horses, equipment and hay; multiple silos; a manager's office; and a fire station. The farm buildings were clustered around a spacious courtyard, and elaborate kennels and gamekeeper's facilities housed hunting dogs and young pheasants bred for the driven game shoots that Field hosted in the grand manner. Cyrus H. McCormick, whose agricultural machinery, which included the McCormick Reaper, revolutionized farming technology, stayed closer to home when he bought a thousand acres outside of Chicago, in suburban Lake Forest. The Gilded Age extended even into the dairy barn, when *Country Life in America* (June 1904) published a photograph of an elegant dinner party being held in a spotless, well-lighted barn, flanked by rows of inquisitive Holsteins peering from their stanchions. It was captioned "A cow-barn fit to dine in."

A Wisconsin Landmark in Iron County *Opposite*
This beautiful round fieldstone barn with interior silo
was designed to offer superior wind resistance
to sudden Midwestern storms. Interior plans for
such a barn usually included a feed room at the base
of the silo and stalls facing the windowed outer wall.
Continuous mangers fronted the stalls, and the
manure alley behind them allowed for greater
ease of cleaning.

Round Banked Barn with Central Silo *Below*
This impeccably kept barn was built by Indiana
farmer George Ramsey in 1910, with masonry stable
walls and a central silo rising through the domelike
roof. The round barn was more difficult to build than
the rectangular, but it offered maximum efficiency in
terms of floor space and feed handling.

Polygonal Barn with Shed-roofed Addition *Opposite*
Octagon Farms, in Wisconsin's Ozaukee County,
uses this turn-of-the-century barn to house its llamas.
Local farmers must have looked twice when they
first saw such an exotic species in the neighborhood.
The fleece of this animal native to Peru has become
increasingly popular for the warmth and softness of
the wool spun from it.

**Gambrel-roofed Barn with
Mechanized Haytrack** *Overleaf*
This Midwestern dairy barn with rooftop lightning
rods kept pace with a variety of twentieth-century
developments, as seen by the hay trolley between loft
and ground level and the capacious metal-roofed silo.

**Round Stone Barn, Indian Head,
Saskatchewan** *Below*
This venerable prairie barn with central silo
incorporates fieldstone, mortar, slats and shingles.
Barn bulding in many styles, both new and
traditional, flourished in Western Canada between
the mid-nineteenth and mid-twentieth centuries.
The region was settled rapidly and became pre-
eminent in grain production.

Multilevel Dairy Complex, Wisconsin *Below*
Twin silos and the cow and calf mural below the
main barn's bow-truss roofline identify this
handsome complex as a dairy farm, located near Port
Washington. The array of lower-pitched additions
and outbuildings reflects steady growth and progress
over the course of decades.

Barns and Silo, York Township, Michigan *Above*
An unusual banked cement silo bound with metal
hoops forms an exclamation point between an
outbuilding raised on blocks and a three-bay barn.
Matching rounded doorways trimmed in white
tie the buildings together visually at serene
Centennial Farm.

153

Prairie Frame Barn *Opposite*

The endless sky of the grasslands rises high above an abandoned two-story barn near Mankota, Saskatchewan. The roof shingles are wearing away in patches, and the boards show signs of warping. Like an old house no longer maintained, the abandoned barn soon begins to deteriorate.

"Top Hat" Prairie Barn, Saskatchewan *Above*

This generously sized gambrel-roofed barn, with unusual dormers lining the upper level, was built in 1917 near Zealandia. Extended roof peaks on the gable ends shelter the loft area. Barns of this size were often used to show and sell livestock at regional gatherings. The twin metal ventilators are crowned by running-horse weathervanes.

Weathered Siding and Concrete *Opposite*
Beginning in the early 1900s, poured-concrete
foundations and silos became popular for their
durability and, in the case of barn floors, ease
of cleaning by hosing down stalls and feed-
preparation areas. This sturdy farmstead is in
Shawano, Wisconsin.

Banked Barn with Gable-end Addition *Above*
A picturesque grey barn with a low addition along
the front has the flared-eave roofline sometimes
called Dutch gambrel. Acres of ripening corn form
the background to this scene from the heartland.

German-style Dairy Barn, Minnesota *Above*
A classic banked barn, with stone stable walls and
a forebay projecting over the stalls, is still home to a
large herd of grazing Holstein-Friesians, turned out
to pasture for the summer months.

Jersey Cow and Nursing Calf *Below*
This gentle short-horned breed, which originated in
the Anglo/French Channel Islands, is usually fawn-
or cream-colored. Elsie, the trademark Borden cow
familiar to generations of Americans, is a Jersey.
Their milk is rich in butterfat, and they are highly
valued as dairy cattle from their native islands to
New Zealand and Canada.

Midwestern Grain Barn *Above*

This windowless frame barn in Wayne County,
Indiana, with its high arched roof and metal
ventilators, is typical of those built at the turn of the
twentieth century. Its cement-block silo is atypically
square, rather than round, the form that would
become the most popular.

A Roomy Michigan Dairy Barn *Below*
Rural Ingham County is the site of this well-kept
barn with silos and outbuildings, bordered by a
colorful phalanx of daylilies.

A Scene Etched in Frost *Opposite*

Late fall brings the promise of respite from the arduous growing season to a Great Plains farmstead framed by gnarled leafless trees feathered in white. The spacious loft under the Dutch gambrel roof of the barn is packed with hay for winter fodder — perhaps timothy, alfalfa, or clover.

Midwestern Monochrome *Above*

Muted shades of grey and black softened by rounded snowdrifts distinguish this heartland complex, which has an old-fashioned tile silo crowned by a metal rooftop ventilator.

Dane County, Wisconsin *Opposite*

As its name suggests, this region was populated during the early 1800s by many immigrants from Scandinavia. Like their contemporaneous Yankee and German neighbors, they sought out land that combined wooded areas for building materials, elevated grasslands that were easily tilled and drained well and low-lying meadowland for pasturage and the cultivation of hay.

Showing Its Colors *Above*

Bold patriotic murals on a deep-blue barn strike a bright note in Kewaunee County, Wisconsin. The state has a long tradition of imaginative barn decoration, revived during the 1970s by the Dairyland Graphics project sponsored by the Wisconsin Art Board and supported by the National Endowment for the Arts.

Room to Grow *Above*

The large tracts acquired by Midwestern and Great Plains settlers allowed for extensions and additions as the farm prospered and family size increased. It was always the barn and its ancillary buildings that received the most time and attention, rather than the farmhouse. As the "factory" of the family business, it was essential to success.

Wisconsin Pastoral *Below*
Simplicity and dignity emanate from this red-and-white complex at Sturgeon Bay, where vertical board siding and a weathertight rounded silo are counterpointed by a star-shaped gable ornament of German inspiration.

Mid-continent: Grassland to Grain *Above*

A weathered prairie barn surveys the fertile plains of
Kansas, rising gradually from east to west and drained
by the Arkansas and Kansas River systems. The contiguous states of Oklahoma, Nebraska and Missouri
share the long growing season and rich soil that have
made this region famous as America's breadbasket.

Dogtrot Barn with Drive-through Passage *Below*
Log barns and cabins chinked with various materials
spread from the Tidewater region to the Southwest, as
seen in this typical example. Forerunners of the ranch-
house style, these barns featured a common roof span-
ning separate cribs, or cabins, to form an open passage
for wagons. Originating in Scandinavia, the style was
adapted to milder climates.

Minnesota Dairy Barn *Above*

The Land of Lakes became a mecca for dairy farmers
from the East and from abroad during the nineteenth
century. This blue-shingled, peaked-roof barn is
typical of those built at the turn of the century to
house and breed growing herds of dairy cattle,
whose milk, cheese and butter could be transported
by rail to urban markets.

Day's End *Overleaf*
Evening envelops a quiet farmstead in Richland
County, Wisconsin, where the many chores of
the day are almost done—for a little while.

**Free-ranging Chickens at
Historic Amana, Iowa** *Below*
Old-fashioned poultrykeeping is part of a bucolic
scene that recalls the original Amana Society, whose
members settled in Iowa in 1855 and founded seven
villages where they lived and worked in common.
The hens take refuge indors by retreating through
the small doorway at lower right during bad weather.

The West

During the late 1840s, a new era of westward expansion was opened up by the annexation of Texas; the conquest of Mexican territory including the present states of California, Utah, Arizona and New Mexico; and the cession of the disputed Oregon Territory (now Washington State, Idaho, Oregon and a former portion of Western Canada) by Great Britain. Seasoned mountain men led wagon trains of American settlers to the fertile Northwest. The mining frontier in the Rockies brought tens of thousands of prospectors, many of whom remained as permanent settlers, establishing ranches and farms in Colorado, Montana, Nevada and Wyoming. The California Gold Rush of 1848–49 increased the American population there from 700 in 1845 to more than 80,000 before 1850, and many who were disappointed in the gold fields turned to agriculture, which flourished in the new state's warm climate and arable valleys.

After the Civil War ended in 1865, many impoverished Southerners, whose agricultural economy had been virtually destroyed, sought to rebuild their lives in the West, as did thousands of former slaves, who settled in both the Great Plains and the Southwest. California and the Pacific Northwest saw an influx of Chinese immigrants who had come to help build the transcontinental railroad and stayed to found businesses and farms. After the annexation of Hawaii in 1898, growing numbers of Japanese fishermen, agriculturists and merchants came to the West Coast, where trade with the Far East was beginning to boom. It was a period of back-breaking labor, high hopes, new opportunities and fortunes made or lost with bewildering rapidity, and as these diverse regions became settled, new forms of vernacular architecture rose to meet the needs of those who were steadily filling the Far West of the United States and Canada.

Previous pages: Jensen Living Historical Farm, Logan, Utah; *Above:* Old barn in western Texas

In the Southwest, indigenous Native American and Hispanic styles had long been combined to protect people and animals from the dry heat of this region, where irrigation was usually required to produce crops. Cattle ranching became a major industry in south Texas, where millions of longhorns descended from cattle brought by Spanish settlers roamed the plains. These animals were very different in appearance from the purebred English longhorn, as seen at Rousham, in Oxfordshire, and now very rare. The English longhorn has horns that curve in sharply toward its eyes, rather than growing outward, and is large and heavy-bodied. However, like the Texas longhorn, it is an unusually hardy animal, living out all winter and highly regarded for the quality of its beef.

As the demand for meat grew in heavily populated urban centers, cowboys of American, African American, Hispanic and Native American descent rounded up the cattle and moved them northward on "the long drive" to railheads including Abilene, Texas, and Dodge City, Kansas. Thence they were shipped to meat-packing houses in Chicago. These sturdy animals required almost no shelter, but the cow ponies used to herd them—captured from wild horse herds and trained by the cowboys to saddle and bridle—were corralled and stabled near ranch houses similar to those originally built by Hispanic and Anglo settlers. Many of these buildings also traced their descent to the Southern "dogtrot" barn, which has a driveway or breezeway running between two log cribs, or cabins.

At the Reynolds-Gentry Ranch (late 1870s), near Albany, Texas, stands a well-constructed wooden barn built of pine boards brought in from the Dallas area by train. On the first floor, grain bins equipped with chutes delivered corn and other feed to mangers along the center of the floor, flanked by horse stalls on one side and storage areas on the other. The stalls were well lit and ventilated by small windows set high in the wall. The second level housed a loft, accessible by a stairway and provided with a hay door for mechanical loading of hay from wagons. The barn was later used by M.G. Gentry to stable his Thoroughbred horses and has been relocated to the Ranching Heritage Center at Texas Tech University in Lubbock.

At Laguna Seco Rancho, near Coyote, California, established during the 1840s, William Fisher built a large frame barn with a hay door in the gable and three wagon entrances. As his grain and stock operation prospered, his son Fiacro enlarged the barn several times, adding a lean-to shed on one side and increasing the height of the building. Part of the earth-floored main level contained numerous stalls, and a plank-floored extension measuring 42 by 63 feet was used for other farm operations. By the late nineteenth century, a three-room office had been built to oversee the business. As with many buildings in the West, the roofline extended to create a sheltering porch on one side, and a gable-roofed porch framed the main entrance.

The J.A. Ranch, near Claude, Texas, addressed the problem of keeping milk and meat fresh with a large rectangular building combining stone walls, lattice-work and a mud-covered roof supported on timber vigas that protruded from the façade. The dairy portion of the single-story building, like the traditional springhouse, had a constant supply of cold running water, and meat was hung from the rafters, as in a walk-in refrigerator. Fieldstone flooring helped maintain a constant temperature around the large water trough.

At the "U Lazy S" Ranch, near Post, Texas, a two-story board-and-batten structure served as a carriage, saddle and harness house. A feed storage room was located at one end, and the second floor was used for general storage. The building had only a few small windows, because there was little need for light and ventilation. Other Southwestern ranches converted early dwellings into summer kitchens, or constructed such kitchens to serve the needs of a growing number of workers. During the hottest months of the year, these separate facilities provided a cool, well-ventilated place to prepare and serve food, much as they had done on prebellum Southern plantations.

In central California, dairy farming prospered as the state's population grew, and large frame barns with wagon drives on either end were used to house the animals. These cow barns usually featured a large central bay, flanked by shed-roofed wings used for storage of feed and implements. The stalls lined each side of the main dairy floor, with its feeding passages.

In *Old Barn Plans*, Richard Rawson illustrates the make-do quality of many Southwestern barns, including one example in Tiptonville, New Mexico. This long, low structure with a gable-roofed main entry at either end is "a patchwork…of log, board-and-batten, adobe, sheet iron and stone." The horse-hitching posts familiar to everyone who grew up watching Western movies line the front of the building.

An unusual 1887 barn has been restored at the Hubbell Trading Post National Historic Site, in Ganado, Arizona, on the Navajo reservation. It is of dry masonry construction, with mud grouting added at a later date. The windowless structure has a slightly arched roof, rather than the flat roofs more frequently seen in this part of the country.

In South Pass City, Wyoming (the "Cowboy State"), is the Black Horse Livery Stable (c. 1868), where horses were hired out or boarded, as needed. A rugged,

nearly square log structure, it is chinked on the outside with cement, and on the interior with wood strips. Finished floor surfaces in the stalls are made of six-inch squared logs, and the adjacent storeroom has a dirt floor. Each stall is equipped with a rough-hewn feeding trough. The construction of the roof trusses resembles the mine shoring used at the nearby Wolverine Lode. This stable is now part of the Old South Pass City Historic Preserve.

Few Western barns have the unusual window and door openings that we associate with the older barns of, say, New England and the Mid-Atlantic states — the curved-spoke round window; star-shaped and other geometric loft openings, or the diamond-shaped "tilt" windows seen in the gable ends of Northeastern buildings. For the most part, these Western barns had few adornments: They were often constructed quickly, from materials newly available, and designed with the

J.W. White Barn (1870), Mason, Texas

emphasis on utility. As on the Great Plains, many featured rain hoods projecting over both of the end gables, especially where these gables were left open to ventilate the hay. The barn and its outbuildings were forthright, unmistakable working buildings.

There were, of course, exceptions, as seen on vast holdings like those of the King Ranch in Texas—a property once larger than the state of Rhode Island. Rebuilt in 1912 after a fire, the complex resembles a Spanish-style fortress. The King Ranch, working with the Texas Experiment Station, undertook extensive work in crossbreeding the American Brahman bull with both Hereford and Shorthorn cows to produce desirable beef cattle resistant to heat and sparse ranges. The great foundation sire of the Brahman breed in the United States was Manso, bred by the Sartwell Brothers of Palacios, Texas. Louisiana and Florida, both major ranching states, have shown increasing interest in such experiments since 1943, when the U.S.D.A. circular "Hybrid Cattle for Sub Tropical Climate" reported that "Hybrid cattle with one-fourth to one-half blood of a Brahman breed and the remainder from a British breed have demonstrated unusual ability to produce beef from grass."

Early in the twentieth century, a portion of the King Ranch at Laramie, Wyoming, worked with state and federal agencies to develop a new breed of sheep for Western ranchers. This was the Columbia, derived from the French Rambouillet crossbred with several types of British ewes, notably the Lincoln. The new breed found rapid acceptance among sheep ranchers because it had a longer-stapled and lighter-shrinking fleece. It also produced stronger lambs, with a high survival rate. According to *Modern Livestock Breeds*: "The Columbia has been bred along lines that appeal to the average sheepman rather than to the person looking for perfection in mutton conformation. Columbias are an extremely large, robust sheep, but are longer of leg and more rangy of body than most of the mutton breeds. The face is open, and the hair on the face and around the feet is white and fine. Hoofs may be either white or black." These animals stand up well to rigorous range conditions like those that prevail from Western Canada through the Rocky Mountain States. The typical sheep-farming barn is designed primarily as a feeding station and lambing center, where newborns are kept with their mothers until they are weaned. Field barns, providing hay and rudimentary shelter, resemble the old-fashioned open-sided sheepcote used in the British Isles, where the climate is far milder than that of the North American West.

Both American and Canadian farmers turned to swine production on a larger scale than had been practiced in the East. Shelters for pigs ranged from a crude shed in a fenced sty on a family farm to a full-fledged barn designed for breeding and rearing pedigreed animals shown at livestock fairs. Originally, pigpens were mainly of the former type, and the animals were allowed to roam and graze, or were fed on waste corn and produce along with kitchen scraps. Farmers who raised pigs on a larger scale built shelters of wood, brick, or stone with adequate ventilation and sanitation.

In 1842 the influential *American Agriculturist* had decreed that "The piggery or hog pen, if not a large establishment, should not be at a great distance [from the farmhouse]....In its rear ought to be a comfortable yard for the hogs to range at proper seasons....I would suggest a general plan of a main entrance at the gable end by a hall running through its entire length, with the stalls or pens on each side, and the swill or feed troughs next to the passage. Overhead, corn or other grain, or various farm products, may be stored." Soon afterward, farm journals were recommending the addition of a cooking room for the feed adjacent to the pens. Wood-fired cookers were vented by sheet-metal flues connected to the chimney, and low doors from the pens gave access to the yards.

At his idyllic Beauty Ranch, in California's Sonoma Valley, the novelist Jack London built a handsome two-story fieldstone piggery surrounded by a grassy walled courtyard. (One newspaper described it as "a Palace Hotel for Pigs.") Once London set his heart on country life, he never looked back. As he wrote to a friend in 1905, "I have just blown myself for 129 acres of land. Also, I have just bought several horses, a colt, a cow, a calf, a plow, harrow, wagon, buggy, etc., to say nothing of chickens, turkeys, pigeons....All this last part was unexpected, and has left me flat broke." According to Clive Aslet, in *The American Country House*, this was only the beginning: "He not only built a dwelling of boulders, Wolf House, which tragically burned before he could occupy it, but erected barns, dammed a stream, terraced hillsides and planted sixty-five thousand trees. The spirit which animated him was not very different from that which persuaded Frank Lloyd Wright to build Taliesin with chicken coop and goat pen attached."

Perhaps London's palatial piggery was influenced by several notable English examples built shortly beforehand. Near Robin Hood's Bay, Yorkshire, Squire Barry of Fyling Hall built a Grecian temple for pigs, complete with portico and pillars. The massive stone base was quarried nearby, and the windows were tapered in the Egyptian style of the Exotic Revival. Reportedly, the squire was so indecisive in his quest for the perfect plan that it took several stonemasons two years to build this porcine monument. According to British historian Lucinda Lambton, the author of *Beastly Buildings: The National Trust Book of Architecture for Animals* (Atlantic Monthly Press, 1985), another memorable pigsty was built by Thomas Durham, a friend of the poet William Cowper, who turned from astronomy to farm architecture on great estates near Badminton, in Gloucestershire. At Swansgrove House, he built a castellated pigsty and stable at a prudent distance from the residence of the Dukes of Beaufort. His famous Castle Barn, visible a mile from the fields, has great square towers at either end, used as dovecotes, and a central cowhouse and barn with massive stepped gables.

Such estate barns were rarely seen in the American West, but notable exceptions include the vast complex built by William Randolph Hearst at San Simeon, California, which he always referred to as "the ranch." Designed by architect Julia Morgan, it rambled through successive wings of the main house, stables, outbuildings and expensive landscapes of a property so large that some of the workmen there had never been off it. Hollywood celebrities and business tycoons received coveted invitations to visit on stationery that bore the imprint "Hearst Camp."

More down-to-earth Western ranchers and farmers benefited from a great many late nineteenth-century developments that allowed smaller numbers of laborers to operate these larger establishments. Mechanical hay carriers mounted on tracks in the loft eased the labor of storing fodder, and manure trolleys and spreaders made field fertilization more efficient. Ventilation improved with the use of wind-driven rooftop rotors that drew air through ducts installed throughout the barn. Flexible stanchions for dairy cows, lightning rods, improved threshing machines and wind-driven pumps with circular sails also came into use. Publications including *Canadian Agriculturist*, *The Farmer's Advocate*, *The Nor'West Farmer* and *Free Press Weekly* contributed to scientific agriculture, even in the most remote areas.

The Canadian Pacific Railway encouraged homesteaders to move West by providing a series of prefabricated farm buildings, with plans, for reasonable prices. Other transportation, lumber and catalogue chains like Sears Roebuck also offered plans and packages ranging from barns and granaries to henhouses, piggeries and implement sheds. According to Bob Hainstock, the author of *Barns of Western Canada*, "The [Canadian Pacific] railway offered a rebate of 50 cents an acre for each acre cleared….Not only was the railway interested in selling its land grants back to farmers, in order to create increased freight traffic to and from Eastern Canada, it frequently sweetened the land deal with its own set of house and barn plans and ready-to-farm packages. Between 1888 and the end of the century, it provided an average loan of $300 per settler for houses and implements, and later offered land on twenty-year leases and loans of up to $2,000 to build barns, houses, or wells."

Industrious newcomers to Western Canada in 1899 included the Doukhobors, a religious sect that grew up among the Russian peasantry during the eighteenth century. Nonconformists to the Russian Orthodox Church, they called themselves "Christians of the Universal Brotherhood." Like the German dissenters and the Quakers, they refused to bear arms, and formed a community near the Sea of Azov. Removed to what is now the country of Georgia, in the Caucasus, in 1840, they flourished until 1887, when universal conscription was introduced. By now some 14,000 strong, they resisted serving in the Imperial armies, and because of the intervention of novelist and social reformer Leo Tolstoy, with other Russian intellectuals, they were permitted to emigrate. Almost half of them settled in Canada, where they were given land in Assiniboia and Saskatchewan. Their leader Verigin, released from exile in Siberia, joined them in 1902 and oversaw development of the Canadian community and smaller settlements in the United States and Mexico. Doukhobor farmers built single-story, gable-roofed barns with sod roofs on board rafters. Constructed of logs, they were often plastered over.

As Bob Hainstock points out, both soil and stone were employed in new ways by resourceful Canadian builders. By his account: "Blood Indian entrepreneurs on the Pothole Coulee Ranch near Lethbridge in 1888 produced a barn unique to Western Canada….Digging deep pits into the soft terrain, and filling the bottom with coal from nearby mines, the Natives then heaped piles of fossilized shellfish onto the burning coal. The coal was allowed to burn for several days, during which time the shellfish were transformed to a kind of cement. The powdered cement was mixed with gravel and sand to provide a form of concrete popular in foundation work. Although no one understood why the shellfish converted to cement, they knew they had enormous supplies to work from."

As the farming frontier advanced all the way to Alaska, an early nineteenth-century innovation became increasingly important. America's first county fair had been held in Pittsfield, Massachusetts, in 1811 under the auspices of gentleman farmer Elkanah Watson. Eager to see European advances in animal husbandry adopted in the New World, he invited local farmers to compete for best livestock. Soon, farm wives and children were involved in sewing, gardening, cooking and other activities that brought rural families together in what would become the social event of their year. Harness racing became a popular feature of these gatherings in the East, where Standardbreds, champion trotters and pacers bred for the ability to trot or pace a mile in the shortest time, attracted crowds of spectators. Hambletonian was the best-known breed sire in the United States, combining the desired features of the English Norfolk Trotter and the Thoroughbred.

Livestock competitions at these popular fairs included purebred swine, cattle, rabbits, poultry, sheep and the powerful draft horses now bred mainly for show—the Clydesdale, Percheron, Belgian and English Shire. Fruits, vegetables, preserves, baked goods, nursery plants and needlework also figure largely in the county and state fair, which remains a major local event to this day. Today it includes many carnival-type amusements, and innovations like the Pig-N-Ford Race in Tillamook, Oregon, where dedicated Model-T drivers negotiate the course grasping a portly—usually rebellious—pig under one arm. It's a far cry from the county fair of a hundred years ago, which opened with a brass band on parade and dozens of yoked oxen groomed to the golden tips of their horns.

Red barn and Ruby Mountains, Elko County, Nevada

Under Herringbone Skies *Opposite*
A weathered frame barn built of local timber on a
cement foundation holds its own in Fremont County,
in eastern Idaho. More than one-third of this state's
land area is wooded with virtually pristine forests.

Anglo Influence, California *Above*
The Hispanic heritage of San Joaquin City was
altered by the mid-century incursion of many
Americans, as seen in this gable-and-hip roofed barn,
whose roof is wearing away. Timber fencing built to
last appears in the foreground.

Waves of Grain *Previous pages*
A winding road through vast wheat fields leads
to a remote ranch in the Montezuma Hills of
California's Delta, a fertile region to the east of
the San Francisco Bay Area.

All-purpose Outbuilding *Below*
The makeshift quality of many Western barns, put
together in haste from local materials, is clear in this
example from Segundo, Colorado, probably used
now as a large storage shed.

Three-part Frame Barn *Above*
The roofing on this tattered barn in New Mexico's
Sangre de Cristo Mountains is giving way to the
elements, and its collapse is now a matter of time.
The untended fencing leans at broken angles.

Designed for Its Site *Overleaf*
This well-made range barn has a roofline that slopes
almost to ground level on one side and a covered
wagon drive on the other. Window openings are
limited to narrow vertical loopholes: Shrinking of
the boards over time admits more light.

Early Twentieth Century *Page 190*
The sheltering roof peak over the hay door on this
Milner, Colorado, barn throws a kite-shaped shadow
on the façade.

Idaho Potato Fields *Page 191*
An old barn with tri-pitched gambrel roof
and missing doors borders a far-reaching
field near Nampa.

Zigzag Fenceline *Opposite*
This efficient method of split-rail fencing was used by pioneers to farm livestock enclosures without the labor of digging postholes. The old red barn, near Boulder, Utah, resembles many of those built from mail-order plans during the late 1800s.

Log Barn and Conestoga Wagon *Below*
Atlantic City, Wyoming, is the site of this sturdy log barn with vertical-plank gable. In the foreground is the ghost of a covered wagon like those that brought many homesteaders to the Wyoming Territory by way of Cheyenne, Cody and South Pass City.

Open Barn Plan *Above*
Three large bays under a steep gabled roof are easily
accessible through movable fencing in this livestock
barn near Drain, Oregon.

A Modest Banked Barn *Below*
Symmetrical windows and a hood projecting over
the hay door from a shingled roof mark a small frame
barn built into an Oregon hillside.

Working Ranch *Opposite*
The Butte Mountain provides a stunning backdrop for this newly re-sided barn in California's scenic community of Mokelumne Hill. Its neat corral and electrified fencing indicate that this farmstead is a flourishing enterprise.

California Cattle Ranch *Above*
Acres of pastureland, and a cluster of whitewashed barns and outbuildings overlooking a still pond, provide a congenial "home on the range" for this livestock operation in Valley Ford.

Roomy Gable-entry Barn *Above*
A gambrel roof of corrugated metal, with flaring eaves, provides ample hay-storage capacity in this weathered barn in Texas. Note the cattle skull above the hay door.

A Western Classic *Below*
Board-and-batten siding, neat white trim and sliding
doors with X-shaped bracing distinguish this
handsome barn overlooking a field of wildflowers
near Llano, Texas.

Scenic British Columbia *Overleaf*
An old log barn commands an incomparable view of British Columbia's inland Okanagan region. Vineyards and orchards border the warm-water lakes here, and vacationers have discovered the natural attractions centered around Vernon, Kelowna and Penticton.

Washington State *Opposite*
Framed by colorful native plants, the spacious barn at Rose Ranch, with its bow-truss roofline, is a landmark in rural South Bend. This region's warm, wet climate produces bumper crops of apples, cranberries, produce and nursery stock, from roses to rhododendrons. Livestock flourishes on the area's well-watered pastures.

Revised and Updated *Below*
Originally built in 1896, this long livestock barn in Orick, California, has been well tended, from its cement foundation to the cupolas on the weather-tight roofline, backed by a rising stand of conifers.

After the Storm *Opposite*
The view from this picturesque barn in Elderwood, California, is all the more breathtaking as a rainbow emerges from the threatening rain clouds.

Livestock Barn, Pacific Coast *Below*
A blossoming almond tree announces the arrival of spring at this farmstead in San Andreas, near the foothills of the Sierra Nevada, California.

Three-story Barn with Extensions *Above*
Mellow afternoon light recalls the glory days of this
Amador County, California, barn, which extends its
spacious wings on either side to fencelines sagging
under the influence of time and neglect.

A Proud Idaho Dowager *Opposite*
A New England-style barn supported on stilts,
with old-fashioned tilt, or diamond-shaped,
windows, is rapidly losing ground to wind and
weather along the Snake River in Owyhee County,
southwestern Idaho.

Historic Tools, Mills and Machinery

It is perhaps 10,000 years since people began to practice agriculture and animal husbandry, and many of the implements we use today are more complex versions of their primitive tools. The first was a pointed stick for digging, originally used by food gatherers to dig roots. The first farmers employed them to make holes for seeds gathered from wild grasses and grains. This evolved into the spade, when a crossbar was added to drive the stick more deeply into the soil with one's foot. Similarly, the first hoe was a stick with a sharp branch at one end; later, a sharp shell or stone was bound to the branch to give it a better cutting edge. The sickle, for reaping hay and grain, developed before the Iron Age, when early farmers set sharp stones along one edge of a curved stick.

Agriculture developed independently in the Americas, where wheat and barley, rice, sorghum and millet were unknown until the advent of European settlers. The principal crop—unique to the Western hemisphere—was maize, called "corn" by the English, who used the word to refer to all kinds of cereal crops.

Potatoes, squash, beans and other foods were also cultivated by Native Americans of various tribes and regions. Because they had never used domestic animals for plowing, they did not adopt a "clean field" agriculture, but planted seeds in hills a few feet apart and cut the weeds only around the hills.

Nearly all prehistoric farmers in arid regions from the Middle East to Central America developed some method of irrigating their crops, whether digging trenches from nearby water sources, or carrying water laboriously by hand in pottery or other vessels. Many also discovered the advantage of using fertilizer, usually manure from domestic animals. In the New World, fish were often buried in hills of maize to increase the yield. Later, the discovery of metal and its uses, marking the end of the Neolithic period, led to the development of stronger and sharper blades for hoes, plow points and sickles, except in the Americas.

Early saws of stone gave way to tools with metal teeth, operated by human, animal, or water power. By the eighteenth century, circular saws were used to cut

Previous pages: A classic windmill, Liberal, Kansas; *Above:* Grain harvest, 1908, at Fallon Flat, Montana

wood in Europe's forests, which were being depleted rapidly even before that time. In the New World, the great stands of virgin forest were both a resource and a trial to pioneer farmers east of the Mississippi, who had to clear their land before planting. However, the timber was put to good use in the construction of dwellings, barns and outbuildings, after it had been seasoned for at least a year.

One method of seasoning was to girdle the bark around the tree trunk with an axe, which caused the tree to die. It was then felled up to a year later, by which time the wood had dried out naturally. Another method was to soak cut logs in water for some time, then leave them to dry in an upright position. The broad axe was used to hew square timbers from logs, and the adze served for rough trimming of the hewn timbers. Chisels of various shapes and sizes were employed to shape the mortises and tenons that fitted structural beams together for post-and-beam construction. The marks of these simple but versatile tools may still be seen on the timbers of historic barns.

Most early barns had a central threshing floor, where the grain was threshed by the time-honored method of flailing by hand. The flail comprised a long wooden handle joined to a movable "swingle" by a leather thong. Lightweight winnowing trays were used to separate the grain from the chaff by tossing the wheat or other grain into the air so that the wind would take the chaff while the heavier kernels of grain fell back into the tray. Later, when the harvesting process became mechanized, the threshing floor would serve as a wagon drive and storage area for bulky implements.

Other tools indispensable to early settlers of North America included the auger, used to make holes in structural beams for joinery; the drawknife, a crude planing tool with handles at each end; and the froe, a dull metal blade with an upright wooden handle. It was placed against the top of a log and struck with a mallet or maul to split clapboards and shingles for roofing and siding. The early three-tined hayfork was often carved from a single piece of hardwood and used to pitch hay from the wagons into the mow for winter feed. Straw was stored separately for livestock bedding. The itinerant master carpenter went from farm to farm to help with barn con-

Wood cutting, near Prescott, Arizona, 1911

struction, using a "story pole" with incised measurements to lay out the framing. Carpenters' marks in the timbers indicated their placement in the structure.

The transformation in European agriculture, especially in England, opened the way to new methods of farming and livestock breeding during the eighteenth century. At that time, former open-field farms were enclosed and much arable land was converted to pasture. This caused great hardship to many small farmers in the British Isles, some of whom emigrated eventually to Canada and the future United States, but a number of leaders in scientific agriculture emerged to transform many traditional practices. Jethro Tull invented a grain drill and advocated more intensive cultivation of English farmlands and the use of animal power. Charles Townshend improved crop rotation and promoted the use of clover as a cover crop to restore nutrients to worn-out fields, much as alfalfa is used today. Turnip cultivation was introduced from Flanders and new grasses from France.

Robert Bakewell was instrumental in developing better breeds of livestock, and Thomas Coke founded a model agricultural estate, working mainly with wheat and sheep. Both Arthur Young and Sir John Sinclair were influential writers, and corresponded with gentleman farmers in the newly formed United States, including George Washington and Thomas Jefferson, who experimented widely on their Virginia estates. Groups like the Philadelphia Society for Promoting Agriculture, founded in 1785, were familiar with the new English practices and helped to promote them. English longhorn cattle were imported as early as 1783, and statesman Henry Clay introduced Hereford beef cattle in 1817. Agricultural journals began to appear in 1810, and John Stuart Skinner first published the widely read *American Farmer* in 1819.

Eli Whitney's invention of the cotton gin in 1793 revolutionized the Southern economy and led to expansion of the plantation system, with its use of slave labor. Many small farmers in the South were marginalized by the dominance of this crop, production of which increased from some 10,000 bales in 1793 to almost 5,500,000 in 1861. Some regional farmers were forced to become tenants of the major landholders, while others made their way across the Appalachians to try their fortunes in the Midwest and the Great Plains.

The development of better plows was essential to the progress of westward migration. In 1793 Thomas Jefferson had developed a new kind of moldboard that offered less soil resistance, but the first patent for a plow was issued to Charles Newbold of New Jersey in 1797. Except for the handles and beam, the plow was made of solid cast iron. Farmers were slow to accept it, believing that the iron would poison the soil.

A major improvement came between 1814 and 1819, when Jethro Wood introduced his cast-iron plow. Its moldboard, share and landside were cast in three parts, and their interchangeability was one of Wood's major contributions to development of the modern plow. His invention was well accepted in New England and the Mid-Atlantic states, but it proved inadequate in the future prairie states, where it would not scour—that is, the heavy soil would cling to the moldboard instead of sliding by and turning over.

John Lane, a Lockport, Illinois, blacksmith solved the problem in 1833, when he began to cover moldboards with strips of saw steel that enabled the sodbusters to cut deeply into the grasslands. In 1837 another blacksmith, John Deere of Grand Detour, Illinois, began to manufacture a one-piece share and moldboard of saw steel. Deere's name became synonymous with the growth of farm technology when his steel and wrought-iron plows displaced the cast-iron model on the Great Plains. By 1850 his farm implement factory in Moline, Illinois, was mass-producing iron plows with English steel moldboards, and he also invented the first riding plow—the Gilpin Sulky—in 1875.

Development of the mechanical reaper was probably the single greatest change in American agriculture between 1800 and the Civil War. The machine made it possible to supplement human power at the critical point in grain farming where the work must be completed quickly or the crop is lost. Since the late 1700s, colonial farmers had been phasing out the sickle and the old-style scythe for cutting grain with the cradle—a scythe with a light framework that gathered the stems and laid the grain down evenly. Many eager American inventors were working toward an animal-powered machine for harvesting grain, the first of which was patented by Obed Hussey in 1833. The following year, Cyrus H. McCormick introduced his reaper, which would dominate the market.

In 1859 came the Marsh harvester, which used a traveling apron to lift the cut grain into a receiving box, where it was bound into bundles. Early in the 1870s, an automatic wire binder was developed, but it was soon superseded by a twine binder. The first threshing machines in North America were imported from Scotland, but an efficient thresher was patented in the United States in 1837. It was closely followed by an improved grain drill, a mowing machine, a disk harrow, a corn planter and the straddle-row cultivator. The commercial fertilizer industry was just getting underway before the Civil War, which initiated an agricultural revolution in the United States during the latter part of the nineteenth century.

The farm machinery widely adopted after the Civil War was primarily horse-drawn, and various devices were used to transmit horse power to such stationary machines as threshers. A number of steam engines had been developed by 1900: Mounted on wheels, they could be moved from one farm to another by draft animals. The first steam-propelled tractors were employed at the same time, but their weight made them unwieldy. Internal-combustion engines, developed in Europe, offered a more practical solution. John Froelich of Iowa built the first workable gasoline tractor of record and improved it to the point where it completed a 50-day threshing run in 1892. However, the change from animal to mechanical power came slowly. Well into the twentieth century, North American farmers stayed with their familiar horse teams, and continued to grow the feed to maintain them. Not until the 1940s, when the manpower crisis became acute during World War II, and the need for farm products reached new heights, did the majority of American farmers adopt the two-cylinder engine. Many of them remember the day when their fathers traded in their teams for the yellow-and-green B-John Deere tractor.

Opposite: Dutch style windmill, Orleans, Massachusetts; *Above:* An Amish farmer's tools, near Kidron, Ohio

Increasing mechanization made many former occupations obsolete. The milkmaid and dairymaid were eventually replaced by milking machines, and the itinerant corn husker and corn shocker were displaced, except on farms including those of the Old Order Amish, who refused to adopt the new ways. Hay was one of North America's principal crops from the earliest times, and required the services of the haymaker, loader, pitcher and wagon driver at harvest time. The tedder turned the hay over in the field to facilitate drying, and the hay presser baled it before their functions were taken over by machines.

Fortunately, many historic mills still stand throughout America and Canada to remind us of the days when wind and water power played a crucial role in agriculture. As Janice Tyrwhitt observes in *The Mill* (McClelland and Stewart, 1976): "Towering against the sky on high ground, a windmill is as magnificent as a square-rigged ship in full sail. A water wheel turning in a stream delights eye and ear like the surge and slow recession of waves on a shore. Worn cogs in immense wooden gear wheels mesh with a cumbersome elegance. In an old grist mill, every surface is wood worn to satin by the gentle abrasion of flour."

The scenic millpond, overhung by trees and vines and dotted with waterfowl, was a favorite gathering place in European and American villages, as farmers waited for the grindstones to turn their grain into flour.

Painters were equally captivated, as we see in John Constable's *Dedham Mill* (1820), painted in Suffolk, England, and Canadian artist Homer Watson's *The Pioneer Mill* (1880), which was purchased for her mother, Queen Victoria, by Princess Louise. The queen, who loved rural scenes and pastimes, promptly commissioned another landscape from the artist, who had never sold a picture before he exhibited this image of his grandfather's sawmill, located in Doon, Ontario.

Conical fieldstone mills lined the shores of Quebec's St. Lawrence River two centuries ago, and shingled windmills on Cape Cod date back to 1793. Long Island, New York, has preserved a handsome Dutch-style windmill at a Bridgehampton park, but the canvas sails have been removed. At the working Prescott Farm Windmill, in Middletown, Rhode Island, each sail carries 180 feet of canvas, set by the miller to face the wind.

Historic gristmills of Canada include the Lang Mill, near Keene, Ontario, which was one of the last in the area to stop grinding. Now a museum, its pond is popular with local swimmers and fishermen. Watson's Mill, in Manitock, Ontario, is an imposing five-story limestone structure designed by master mason Thomas Langrell and millwright Owen O'Conner. The millwright left his stamp upon the complex inner workings of colonial mills much as the master carpenter did on his timber-framed barns.

Landmarks of Atlantic Canada include two mills owned by the Ives family of Prince Edward Island. The grist mill at North Tryon received the island's first flour rollers in 1902, when Charles Ives attracted customers from as far away as Charlottetown. The adjacent sawmill was owned by his brother George. Near Tatmagouche, Nova Scotia, Alexander McKay built the Balmoral Grist Mill to utilize the water power of Matheson's Brook.

Colonial outposts on the Atlantic Seaboard depended on their early mills to free them from the laborious work of grinding grain and hewing wood by hand. The Wolf Pen Mill rose in Jefferson County, Kentucky, near present-day Louisville, when this was part of the first frontier. Its brick-and-stone foundation is two feet thick and laid up without mortar. A 26-foot wooden wheel powered the grindstones.

The Mid-Atlantic states have preserved many beautiful mills, including the Old Red Mill on the Raritan River, in Clinton, New Jersey, and the restored gristmill at the iron-working village of Batsto, New Jersey, which has a projecting bay from which to lift grain sacks from ground level to the loft. In Pennsylvania, the pre-Revolutionary Haines Mill on Cedar Creek was rebuilt to its original five-story eminence after a fire in 1905. A gambrel roof is crowned by a large cupola and lined with dormer windows to light the attic.

As Janice Tyrwhitt points out: "Whether they grind wheat flour, oatmeal or corn, grist mills have attracted most of the legends of milling. Perhaps because they satisfy our nostalgia for the past, more have been preserved than any other type of mill. Few restored sawmills or textile mills exist outside museums and pioneer villages….Every run of grindstones has its own voice." The appeal of these time-honored buildings was captured by the poet Robert Louis Stevenson in the stanzas of "Keepsake Mill," which recalls stolen boyhood visits to the mill near his home in Scotland:

Here is the mill with the humming of thunder,
Here is the weir with the wonder of foam,
Here is the sluice with the race running under—
Marvellous places, though handy to home!

Sounds of the village grow stiller and stiller,
Stiller the note of the birds on the hill;
Dusty and dim are the eyes of the miller,
Deaf are his ears with the moil of the mill.

Years may go by, and the wheel in the river
Wheel as it wheels for us, children, to-day,
Wheel and keep roaring and foaming for ever
Long after all of the boys are away.

Opposite: Gears and pulleys on an old combine, California Expo; *Above:* Grain elevator in Craigmont, Idaho

Windmill and Cistern *Above*
Still in use in Lancaster, Pennsylvania's, Amish country is the American windmill, invented by Daniel Halladay in 1854. Only a breeze is needed to power the mechanism, which pumps water into the adjacent cistern.

Western Sentinel *Opposite*
The arid Southwest was settled and farmed with the help of the tower-type windmill, with its revolving annular blades and rudder, as seen in this peaceful setting along the old Santa Fe Trail near Greenville, New Mexico.

Tack Wall, Kidron, Ohio *Above*
History is ongoing in this Amish barn, where
harnesses, reins and collars for the patient draft
horses are still essential for cultivating fields in
which the gasoline-powered tractor has never
been adopted.

Sugar House, Reading, Vermont *Page 220*
The tall flue rising through the steeply pitched roof
of this old sugar house vents the steam produced by
the slow process of boiling spring's maple-sugar tree
sap into syrup.

Broom Straw and Hardwood *Opposite*
Handmade implements as used in the colonial era
include this rustic broom and hay fork, which is
carved from a single piece of hardwood and
displayed at Philipsburg Manor, once a 90,000-acre
estate owned by prosperous Hudson Valley flour
merchant Frederick Philips.

Water Wheel *Page 221*
This venerable colonial gristmill (c. 1720), built
on the Pocantico River in present-day North
Tarrytown, New York, has been restored at historic
Philipsburg Manor under the auspices of master
millwright Charles Howell, a fifth-generation
miller from England.

Feed Chute and Manger *Below*
Hand-carved mangers for livestock were often
supplied by convenient chutes descending from
a bin or barrel where feed was stored in quantity
for ease of handling. Some mangers were made
for single stalls; others were continuous, as in
the sheep barn.

Drive-floor Storage Area *Opposite*
Wide gaps between the barn siding help to illuminate
a central drive floor used to store farm implements,
wood and drying foodstuffs. When threshing
became mechanized, the former threshing floor
was often used mainly for storage of bulky or
seasonal equipment.

Early Riding Tractor *Opposite*
Nineteenth-century advances in agriculture included many horse-drawn implements that would later be powered by machines, like this iron-wheeled tractor on which the operator could sit instead of standing. Such inventions gained wider acceptance after the Civil War, which created a great demand for farm products and a concomitant shortage of labor.

Stepping Proudly *Above*
A handsome team of draft horses guided by a farmer in Little Sands, Prince Edward Island, shows how harvesting was carried out by manpower with the aid of animal power—a condition that prevailed on many small farms until World War II, after which North American agriculture was commercialized as never before.

A Traditional Hay Rake *Opposite*
Historically one of the continent's three principal crops, hay became more profitable with the introduction of implements like this high-wheeled metal rake, first produced during the mid-1800s. Other major developments included the McCormick reaper and the John Deere steel plow.

Five-horse Hitch for Disk Harrowing *Above*
Five horsepower and a traditional harrow is all this Amish farmer needs to break ground for sowing his Ohio field for the next growing season. In the distance is his prosperous farmstead, which has expanded widely since its establishment during the 1800s.

Details and Ornamentation

Previous pages: **Painted wall and windows in Waitsfield, Vermont;** *Above:* **Interior framing of a round barn from 1911, Bridgeman, Michigan**

The simplicity of vernacular architecture is one of its greatest attractions, but old barns often display handcrafted details and ornaments that contribute to their beauty and originality. Some of these features are unintentional, like the tool marks on seasoned timbers, or the skillful interlocking of fieldstones without mortar in a sturdy foundation. The master carpenters and masons who created the buildings took satisfaction in a job well done, while admirers from a later period see the unmistakable mark of handcraftsmanship as opposed to impersonal machine-made materials and technology. A concrete silo is not an object of beauty to the average beholder; a weathered corn crib or an old-fashioned hay barrack seems to tell us something about the people who made and used it.

Early North American farmers followed the principles outlined in *The Carpenter's Pocket Dictionary*, published in 1797: "Strength and convenience are the two most essential requirements in building; the due proportion and correspondence of parts constituting a beauty that always first attracts the eye; and where that beauty is wanting, carving and gilding only excite disgust." Of course, there was little possibility of carving and gilding a barn constructed from a clearing hacked in the forest to shelter a hard-working family and its animals. But the primitive log cribs and clapboard barns of the first frontier have a natural beauty that appeals to us in a world that seems to become more complex at every turn. Perhaps this accounts for the renewed popularity of log cabins, country cottages and other rustic retreats with no pretensions to grandeur, built along lakes and streams in out-of-the-way places as weekend or retirement homes.

Certainly, the contemporary concern with preservation, conservation and ecology speak to a deeply felt need to reconnect with our collective past. So does the burgeoning interest in folk art and crafts, as practiced by generations of Americans and Canadians. Such folk art appears on many old barns in the form of "hex signs," paintings of animals, landscapes, contrasting colors, ornamental datestones, dovecotes and swallow holes in the eaves. Immigrants from Switzerland and central Europe brought the custom of painting their barns in bright blues, reds and yellows that picked out the exposed beams and rafters. The connected barn, which originated in the *Loshoes* or house/barn, has distinctive rooflines and window patterns in its various sections that indicate their uses: dwelling, dairy, tool-shed, byre, stable and so forth. After the Revolutionary War, patriotic themes became popular, and barns were sometimes adorned with legends including "1776" and paintings of the Stars and Stripes. Even the unmistakable profile of George Washington appeared on barn siding during the nineteenth century.

Many old barns in Quebec and Atlantic Canada have wagon doors and window trim painted in brilliant colors, some bearing the emblem of a dairy cow, a draft horse, or another animal associated with a particular farm. Other Eastern Canadian barns were whitewashed with a mixture of lime and water, sometimes combined with a pure grade of ground chalk. The flared bellcast roofline and the occasional mansard roof speak to the region's French heritage, which is also reflected in its folklore. Tales that combined the everyday and the fantastic were told by the *habitants* when they gathered for amusements like card-playing and storytelling, as recalled by W. P. Greenough in 1897:

> *In one of Nazaire's stories of an enchanted princess, the princess's deliverer, Petit Jean, finds in the enchanted palace a table spread with smoking hot viands, among which were boiled pork, sausages, and other delicacies held dear to the Canadians. The liquors were whiskey and rum, the latter being the best real Old Jamaica. In a larger and grander hall was a still superior table which furnished not only these, but patés and black puddings, with wines and brandies, the latter being the best French brandy. In the stable were horses and carriages in great numbers, a "beau petit buggy" being among the vehicles.*

Trotters and pacers were as popular in Eastern Canada as they were in New England and New York State, where harness racing was followed avidly (and still is). These horses often appear on barn weathervanes, as do dairy cows, sheep, pigs and, especially, roosters — the origin of the synonym "weathercock."

In seaside communities, whales and fish were popular rooftop ornaments. Other traditional weathervanes, crafted by the blacksmith or the farmer himself, took the form of a pointing hand, an arrow, or another motif with special meaning to the region. These ornaments served the useful purpose of predicting the weather on the basis of prevailing winds, which the farmer could interpret as well as the sailor and the shepherd: It was essential to their livelihoods. Farmers also consulted the annual almanacs, which provided "prognostications" based on the relative positions of the celestial bodies. The actions of plants and animals, many of which are extremely sensitive to changes in barometric pressure, humidity and other conditions, were another guideline to the best times for planting, harvesting, or taking shelter from an impending storm.

Masterful stone- and brickwork are often seen in barns, silos and granaries built by the Dutch, Germans, Scots and members of the religious communes, including the Shakers and Mennonites. As Richard Rawson points out in *Old Barn Plans*: "The Shakers were blessed with a willing and highly productive labor force, which enabled them to build such ambitious projects as the round barn [at Hancock, Massachusetts] and the great bank barns on their several principal communes in Massachusetts, New Hampshire, Ohio and New York.

The familiar dairy cow, Summerfield, Prince Edward Island

Patriotic painted decoration adorns a barn in Traverse, Michigan

One sees in both the more traditional stone and wooden barns a Pennsylvania barn writ large. Built into a hillside, Shaker bank barns may have as many as three ground-level entrances and are commodious enough to house all the farm operations that can be performed indoors." Dedicated Shaker craftsmanship extended to the tools, furniture and household objects they produced for their own use and for sale, which have become collectors' items.

Both the Amish and the Mennonites still produce the two-wheeled buggies they use for transportation instead of automobiles. Harness-making and wagon repair are carried out in conjunction with their farming activities. The number of dairy cows on the typical Amish farmstead has increased from half a dozen at the turn of the century to herds of more than thirty animals, and sheep-raising remains an important activity.

Pioneer farmsteads in French Canada were often enclosed in log palisades for protection against wild animal attacks and native war parties, as were English and Dutch settlements of the early seventeenth century. As people migrated westward, they often used tree roots for fencing and crude livestock shelters, or built the zigzag fences that did not require laborious digging of post holes. Later, split rails and wooden pickets were installed where wood was available, while New Englanders used the stones cleared from their fields to build boundaries. During the nineteenth century, gentleman farmers became more numerous and more ambitious. A retired stockbroker named Thomas W. Lawson built the estate he called Dreamwold in Egypt, Massachusetts, with the help of Boston architects Coolidge and Carson. A varied complex of grey-shingled, gambrel-roofed barns and outbuildings, it was surrounded by fourteen miles of picket fence covered by picturesque rambling roses.

New England's historic barns and outbuildings became influential far from the East Coast, as westward migration changed national demographics. The Ellis Stone Barn in Wellesley, Massachusetts; the hillside barn at the John Cram Farmstead in Hampton Falls, New Hampshire; and the Governor Jonathan Belcher Place in Milton, Massachusetts, all dating from the eighteenth century, served as models well into the nineteenth. Meanwhile, the growing prosperity of the Victorian era, and the enthusiasm for wholesome country living among the rising middle class, brought a new look to rural America. Decorative gazebos, carriage houses, kennels and greenhouses began to appear in suburban areas, and estate farms combined traditional and decorative forms in their architecture. At the Dreamwold estate mentioned above, according to Clive Aslet, "There are barns for stallions, brood mares, farm horses and carriage horses; also a foaling barn, stallion service building, [veterinary] hospital, racetrack and riding academy. Equal thought has been given to the needs of the Jersey herd. The cow barn is in the form of a deep, south-facing U, so that the gutters dry quickly. In front of the cows, who stand two to a stall for company, is a track wide enough for the wagon to pass down as farm hands fork feed into the mangers....Two thousand hens occupy a show hen house."

In 1867 *The American Agriculturist* suggested improvements to the traditional three-bay barn and published a sketch entitled "A Good Farm Barn" for changing times. It included stabling for six cows and four horses on one side of the drive floor and a two-story haymow on the other. An attached lean-to shed with a sheltered area in front could house additional animals and a steam boiler for cooking feed. A small door in the gable was used to load corn and feed, and the main entrance, on rollers, had a small "pass door" inset on the right side.

Colorful hex signs on an Albany, Pennsylvania, barn

Above: Kidron, Ohio; *Opposite:* Grand Manan Island, New Brunswick

and some rooflines had as many as three or four cupolas, or a single large one that could be reached by a stairway and used as an observatory.

Until the automobile and the tractor displaced the horse, most towns had a livery stable with horses for hire, and stables were as familiar as garages are today. Every farmer had a "horse story," and inexperienced handlers risked being kicked, run away with, or pinned against the wall by a reluctant mount. One nineteenth-century country journalist reported that, "My grandfather owned a little white horse that would get up from a meal at Delmonico's to kick the President of the United States....I had my usual trouble with him that day. He tried to jump over me, and push me down in a mud hole, and finally got up on his hind legs and came waltzing after me with facilities enough to convert me into hash, but I turned and just made for that fence with all the agony a prospect of instant death could crowd into me." Experienced horsemen still chuckle over the "dude" who doesn't know his way around the paddock, whether on a Thoroughbred farm in Kentucky or a Western ranch like the one depicted in the popular film comedy *City Slickers.*

Even the humble toolshed was targeted for improvement by nineteenth-century farm journals and magazines on country life. One example, published in *The American Agriculturist* in 1862, was an attractive board-and-batten structure with ornamental bracing, a bracketed hip roof and a neat cupola. It was 20 feet long and 12 feet wide, and one interior wall had a diagram of how the tools should be arranged for efficiency: "All ranged in order, and disposed with grace/Shape marked of each, and each one in its place."

Such precision was rare in rural North America, but where the critical eye perceived hard labor, rude living conditions and isolation as the farmer's lot, he and his family saw the humble beauty of their dwellings, barns, fields and orchards and the abundance expressed by a legend discovered on the datestone of an old barn in Guelph, Ontario:

When your barn is well filled, all snug and secure,
Be thankful to God and remember the poor.

A root cellar was located below the right-hand bay. By this time, the ornamental cupola had become a fixture on American barns. Originally designed as ventilators, they became increasingly decorative, with louvers or windows on all four sides and small gabled or hipped roofs, often crowned by a weathervane. Pigeonholes, carved designs, monograms and finials proliferated,

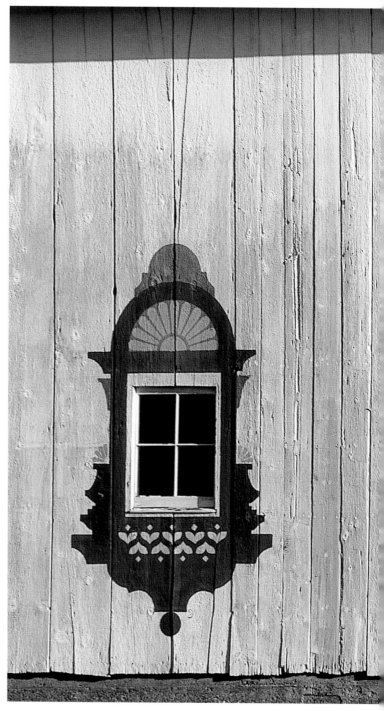

Pennsylvania Dutch Designs *Opposite*
Traditional "hex signs" on a pair of Berks County, Pennsylvania, barns show the circle, star, leaf and flower forms that are also found in regional quilts, samplers and other examples of folk art. Long supposed to be charms against witchcraft, these barn paintings have now been shown to have a purely ornamental purpose.

Bavarian Inspiration *Above*
The delightful Zehner painted barn in Saginaw County, Michigan, reflects the Bavarian custom of decorating house/barn doors and windows in bright colors and detailed designs. This custom also prevailed in Switzerland, and both Swiss- and German-style barns were popularly called "Switzer," a corruption of *Schweizer*, meaning "Swiss."

Barn Mural, Irving Ranch, California *Opposite*
This El Dorado County barn is a burst of color in the landscape, with its gold and green painting of native sunflowers in full bloom. In the foreground is a time-worn disk cultivator.

Proudly Canadian *Overleaf*
A Viking ship and patriotic maple leaves proclaim both the Scandinavian heritage and the current loyalties of this farmstead in New Denmark, New Brunswick.

Midwestern-size Canvas *Below*
This southern Michigan barn became a landmark when its gable end was painted with a pensive, larger-than-life portrait after the Old Masters, distinctively "framed" by the shape of the gambrel-roofed structure.

Smokehouse, Grand Manan Island *Above*
Eastern Canada has a long history of brightly
painted barns and outbuildings, as seen on this
colorful frame-and-shingle fish shed at Seal Cove,
New Brunswick.

A Maritime Motif *Opposite*
Prince Edward Island's Beach Point, one of Atlantic
Canada's many maritime communities, is the site of
this green-and-white smokehouse with a fish tail
fixed to the transom.

The Mason's Handiwork *Right and below*
Fieldstone laid up with mortar made on the site
has provided weatherproof shelter for generations
in the two-story Pennsylvania barn at right, with
its unusual double doors on rollers. Below, a brick
arch and fieldstone walling are revealed by
crumbling stucco facing on an old New Jersey barn.

Rural Kent, Connecticut *Opposite*
Red, white and gold form an autumn montage at this New England farm, with its pass door marked by a cattle skull that evokes the Far West. Horseshoes, for luck, and crossed sheaves of grain are among the other traditional ornaments over barn doors.

Hand-wrought Hardware *Left and below*
Iron monograms and braces forged by the colonial blacksmith were often used to strengthen stone walls, which expand and contract with changing temperatures. Iron hardware was also used to secure doors, as nails and for wall storage—both hooks and baskets. The example at left is from Willow Street, Pennsylvania. The heart-shaped lock and door hook below have been preserved at Rowayton, Connecticut's, Pinkney Park.

Gothic-style Gables and Cupolas *Opposite*
Nineteenth-century American Gothic is exemplified
in the louvered cupolas and gable ornaments of these
weathered barns in Highspire, Pennsylvania. The
finials that crown the bracketed cupolas were fitted
with lightning rods.

Ornamental and Useful *Above and right*
The octagonal louvered cupola above, with its
traditional running-horse weathervane, provides
light, ventilation and decoration on a hip-roofed barn
in Morristown, New Jersey. At right, a large square
cupola, typical of Vermont barns, has small gables
embossed with the letter "K," the emblem of
Kenyon's Farm, in Waitsfield.

Elegant Folk Art *Opposite*
The colorful rooster, pre-eminent among weathervane ornaments, crowns the handsome New England barn at Casey Farm in Saunderstown, Rhode Island. The bird's popularity is reflected in the synonym "weathercock" for the farmer's original aid to weather forecasting.

Weather Sentinel *Above*
A hand-crafted angel with trumpet, the symbolic messenger, communicates the wind direction and watches over the farmstead on this Wilton, Connecticut, rooftop.

Weathervane at the College of Agriculture, University of Wisconsin *Right*
Appropriately, a cow weathervane surmounts this large ventilator on the historic dairy barn at the university's Madison campus. Built in 1898 at a cost of $16,000, the huge three-story building, based on Norman prototypes, was designed by Chicago architect J.T.W. Jennings.

Bold New Brunswick Profiles *Above*
Gable-roofed smokehouses, for preserving the
fishing fleet's harvest, line the waterfront at
Cape Pele, defining the coastal province where both
farmers and fishermen still live by nature's bounty.

Geometry of the Great Plains *Opposite*
This massive grain elevator in Shaunavan,
Saskatchewan, a study in angles and planes, is a
monument to the time when these structures housed
tons of wheat along the transcontinental rail lines to
provide the staff of life for two nations.